Glittering Prospects

Glittering Prospects

ALL YOU NEED TO KNOW ABOUT 'TREASURE-HUNTING'

JOAN ALLEN

PRODUCED BY ELM TREE BOOKS
FOR JOAN ALLEN ELECTRONICS LTD.

First published in Great Britain 1975
by Elm Tree Books Ltd
90 Great Russell Street, London WC1

SBN 241 89090 X

Illustrations by David Clark Design Group

Printed in Great Britain by
Richard Clay (The Chaucer Press) Ltd
Bungay, Suffolk

Contents

List of Illustrations vii

Introduction 1

1. Origins and Growth of the Hobby of Treasure-Hunting 5
2. Equipment, Code of Conduct and Treasure Trove 10
3. Principles and Characteristics of Metal Detectors 22
4. How to Search Houses and Gardens for Hoards and Small Finds 35
5. Descriptions and Locations of Hoards Found in Britain 44
6. Searching for Hoards and Other Objects in Commons, Fields, Farmland, Forests, Ponds and on Footpaths 54
7. Rivers and Tidal Estuaries: the Treasure-Hunter's Most Productive Sites 72
8. Beachcombing: How to Locate and Prospect on the Right Beaches 87
9. A Selection of British Beaches: Where and When to Search Them 102
10. Bottle-Collecting 113
11. Portraits of Three English Treasure-Hunters 119
12. Archaeologists and Treasure-Hunters: Their Strained Relations and Ways to Solve Them 132

Index 143

List of Illustrations

Colour plates

1. Professional treasure-hunter, John Webb, rakes the Thames near Blackfriars—a most productive site.

2. Another professional, Tony Hammond, searches a favourite Sussex beach with his two daughters. (*Photograph courtesy of The News Centre, Portsmouth.*)

3. Antique bottles. At left of upper shelf are Codd's bottles with stoppers; below are ginger-beer and hot-water bottles.

4. Antique bottles come in hundreds of shapes and sizes; the rarest fetch £60 and upwards.

5. Two handsful of coins: Roman, medieval and modern. This lot is worth more than £1,000.

6. The author admires a Unite, a 24-carat coin dated 1615. In good condition it would be worth £300 or more.

7. Old-style pot lids are in big demand. Here is a selection.

8. An experienced prospector found all these coins and bottles in one day's searching. The red object is an underwater metal detector.

Half-tone plates

1. The first step. My sales manager, Peter Bettis, describing the features of a BFO machine to a customer.

2. The thirteen rings on these fingers were all found under the sands of Littlehampton beach in Sussex. Thousands more rings lie under British beaches.

3. It's a hobby for kids as well. The curious youngsters testing these BFO units can learn within hours how to use them well.

4. This lass hasn't yet worked out the right angle for her BFO (the head shouldn't tilt) but her concentration is fine.

5. Shrewd application. This young hoard-seeker stands on a wooden block to stretch the detector head as far as possible into the tree.

6. Good technique with an IB machine. The user's whole mind is on the job; and the detector head as close to the ground as possible.

7. John Webb searches a foreshore with an IB between firmly secured line and pins.

8. John Webb again, seemingly a little distracted as he sieves in the Thames.

9. Pubs old and new are fine for treasure-hunting. People gather in large numbers and, not surprisingly, become careless.

10. Tony Hammond using one of the best coinshooting machines; a PI unit with rechargeable batteries.

11. Signals in high-frequency detectors can drift. A dedicated amateur, Tony White, here adjusts the tuning of his IB unit.

12. Tony White uses his trowel to dig up a find. Note that he has not laid the detector on the ground as he digs.

13. The PI machine Peter Bettis has here can detect a tenpence piece in wet ground down to fifteen inches.

14. The BFO seen here has a six-inch detection range, shallow enough for the searcher to use a screwdriver for retrieval.

15. Large hoards have been found in derelict buildings. No place is too unlikely, even a flimsy wall like this.

16. More knowhow in searching an old house. Lintels and doorframes must always be carefully checked.

INTRODUCTION

The pastime of treasure-hunting began to develop quietly in Britain in the late 1960s. For the first couple of years the growth was sedate. This was inevitable. The modern treasure-hunter is a person equipped with an inexpensive, but sophisticated, device called a metal detector. Around 1968, when the hobby started to spread, there were few detectors to be had at a reasonable price. Imported ones were naturally expensive. Only a trickle of low-priced detectors made in Britain were coming on to the market.

Demand was certainly there but domestic production was literally a cottage industry. The earliest metal detectors were almost entirely beat-frequency machines that a few enthusiasts were assembling in their kitchens, garages and woodsheds. Their makers were fiercely keen and, indeed, some of them were highly qualified in the field of electronics. But, overall, the assembly of machines itself was little more than a pastime, a moonlighting activity that people were undertaking in their evenings and at week-ends. So production was slow; a hard-working maker could assemble and complete perhaps eight machines a week.

Then, in 1970, a few programmes about treasure-hunting opened the floodgates of public interest. Overnight tens of thousands of people were jerked into awareness of the extraordinary volume and variety of articles to be found. A great many of these fascinated people around the country wanted to know where they could lay their hands on a metal detector which would assist them to search wherever they thought something valuable or interesting might be buried. Perhaps coincidentally, one or two makers decided to take the plunge and expand their so-far meagre output; an area of production moved from the kitchen-table to the modest assembly-line.

As the flow of machines was stepped up, more publicity followed, in newspapers and magazines and on television; treasure-hunting was really on its way. By 1973, an estimated 30,000, or more, people were active in widely varying degrees of interest and competence. By this time, I had been selling metal detectors for about three years. This was my second business venture, but it certainly didn't come about by shrewd anticipation on my part. I had not carefully and nimbly leapt aboard a bandwagon I had sensed was about to start rolling. In fact, it was an indirect, almost haphazard, process that developed because metal detectors had a link with my first business, as a coin dealer.

This connection, which was obvious enough, could have prompted somebody to mail me a leaflet about metal detectors, with a letter asking if I would buy some to sell. I thought it would be a good idea to have one in the showroom; I was always looking for interesting things for coin clients to see. The machine arrived and, almost immediately, a client bought it. Then I ordered six more. These sold quickly; my next order was for twenty-four machines and suddenly I was in the metal detector retail business. For personal reasons, I decided to move out of the coin dealership to concentrate on metal detectors. It was a coincidence that treasure-hunting and, therefore, metal-detector sales, began to flourish not long afterwards.

It was through the same casual approach, and unsensed matching growth in public interest, that I became a coin dealer back in 1964. I had given up a good secretarial job because of a complicated slipped disc. Needing a hobby, I thought about coins and took a few Victorian bits and pieces from my father's modest collection to Spink, where they were graciously priced at three or four pounds. This intrigued me, so I put a small—I think it cost about seven shillings—advertisement in *Exchange & Mart*. To my amazement, little registered parcels started arriving in the post. From that moment, my hobby had become a small business: I was a coin dealer.

I then bought a few catalogues and coin books from B. A. Seaby, who had become my good friends through regular contact. By trial and error, I learned the basics of my tiny business, which was now fetching me a few pounds a week. Luck was very much on my side. For, at precisely that time, an intense collecting boom was building up. People were buying coins, Georgian silver, stamps and whatever they could find as investments. My business tended to focus on these individual investors, rather than with other coin dealers.

I used to build up collections of George VI and Elizabeth II coins for clients. I was very meticulous, refusing any coin that was not perfect, in absolutely mint condition. This gave me an edge over some excellent dealers whose notion of mint-condition coins was not as painstakingly precise as my own. Care bred trust and the right coins came to me. I was able to assemble collections of the two reigns I mentioned above and sell them to investors from £1,400·00 upwards.

It is now almost impossible to acquire mint coins for these collections: pieces like 1959 florins, half-crowns, 1959 Scottish shillings in perfect mint state have vanished from the market, although plenty of these coins just below perfect still circulate. This means that these unflawed collections have soared in value. By 1980, they should be worth at least £25,000 or perhaps much more—a brilliant anti-inflation investment. (I should state that by mint condition, we mean coins as struck by the Royal Mint with no Bag markings.)

Early in 1974, I put on a third business hat, this time as a dealer in antique bottles and pot lids. This move was somewhat more calculated than the first two business ventures. My four years selling metal detectors had given me an insight into collecting crazes. I sensed that antique bottle collection could become very big indeed. It had so much going for it: the abundance, the charm of old bottles and a nostalgia that was practical because the items were within a comfortable price range for thousands of people.

Finally, a few words about this book. As a widespread pastime, treasure-hunting (or prospecting as it can often be accurately called) is still in its infancy, so instructional literature about it has been very sparse. This book does not profess to be an encyclopaedia on the subject. But, as the subtitle states, it does tell newcomers all they need to know to get themselves properly embarked as treasure-hunters. It also contains tips on research, selecting sites and actual metal-detecting techniques that a great many treasure-hunters of moderate experience do not know, because they have not had access to effective guidance. For the first time, the book sets out the direct views of archaeologists on the thorny subject of their relations with, and coolness towards, treasure-hunters. It sets down their advice on how tensions between the two groups can be eased—an important topic for both sides.

Many people helped, and I am grateful to them all. Special thanks go to John Webb for his counsel and the considerable time he devoted to the subject; Tony White; Tony Hammond for valuable information and the loan of the superb finds that adorn the book's cover; my sales director, Peter Bettis, whose encouragement was vital; David Clark Design Group for the illustrations; editorial advisers at Elm Tree Books; and John Sheppard for his help and patience in helping me put the manuscript together. Grateful thanks also go to these companies and their executives: Best Electronics Ltd.; Richard Dodds; C-Scope Co. Ltd.; Jerry

Schneider and Martin Fry; Fisher of America; Alan Ogilvie; Geo Electronics Ltd.; Eric Foster; Roderick and Gregory Young of Young Electronics.

Alan Biddlecombe helped considerably on the bottle scene, and I also salute the efforts of my growing number of agents throughout the country.

My hope is that treasure-hunters/prospectors who read this book find themselves more enlightened, more dedicated to responsible and common-sense pursuit of the hobby, and richer—in that order. I can't really ask for more.

No doubt when you have finished reading this book, you will be firmly hooked on the hobby of treasure-hunting or bottle-collecting, and should you require regular information by way of magazines, my company does in fact market a regular treasure-hunting magazine entitled *Scan* which incorporates features, photographs, competitions, etc. on both treasure-hunting and bottle-collecting.

We also issue regular Bottle and Pot Lid Bulletins which keep the collector well abreast of the value of their collections.

Should you require any further information regarding these publications, just drop us a line at our head office at 184 Main Road, Biggin Hill, Kent.

JOAN ALLEN ELECTRONICS LTD.
Metal/Mineral Detector Retailer and Consultant
184 MAIN ROAD, BIGGIN HILL, KENT, ENGLAND
Telephone: Biggin Hill 4234 (01 269 4234)

CHAPTER 1

ORIGINS AND GROWTH OF THE HOBBY OF TREASURE-HUNTING

AMATEUR treasure-hunting yields results only when done with a completely professional approach. Although Britain is replete with buried valuables, searching for them with a metal detector is not a short-cut to quick and easy riches. Achieving success in this pastime is like succeeding in any other area of life: the people who find the most, and the best, items are those who put the most intelligent thought and effort into the work. Patience is possibly the most important ingredient for treasure-hunting, which has been one of the fastest-rising hobbies in Britain over the past decade. Without it, you are ineffectual. You cancel out the benefits of the most diligent research, the shrewdest selection of working sites and the finest equipment on the market.

Good fishermen are patient. They have developed the ability to suppress their sense of excitement, to approach the job-in-hand in a quiet, self-contained way. Perhaps this is one reason why so many British amateur treasure-hunters (there were an estimated 35,000 altogether in Britain in 1974) have come to it from angling, and why they achieve such good results at their new pastime.

The metal detector, key to the swift growth of the hobby since the late 1960s, has, in fact, been described as a fishing-rod for treasure-hunters. The two activities literally merge when the treasure-hunter uses his machine in river, pond or along a sea-shore. He has become an electronic fisherman, seeking coins instead of cod, brooches and buttons instead of bass and bream.

People who have never done it, or have never tried to find out what most of the practitioners actually do, have fuzzy, even lurid notions of treasure-hunting. The simplest thing is to reach for a cliché. So, to outsiders, it often conjures up childhood memories of Robert Louis Stevenson's superb adventure story *Treasure Island* and those pirate movies in which Errol Flynn inevitably skewered Basil Rathbone through the brisket in the final reel.

Such mental images are colourful and exciting, but they have very little to do with the hobby I am explaining in this book. For most of the people most of the time, treasure-hunting is prosaic, demanding and tinged with slight discomforts. Of course, the excitement of anticipation is always there. You can scarcely fail to find interesting, and often valuable articles every time you work an industrial river foreshore or popular resort beach.

However, the path to that moment of discovery can be arduous: long hours of research poring through books, magazines and studying maps; more time spent studying tidal movements at rivers and beaches; fruitless visits to sites when inexperience or misleading information will thwart your efforts. You will need some luck; few accomplishments are ever made without it. If you prepare soundly and apply yourself sensibly you will need smaller doses of luck to achieve your successes.

That last point is a truism. Yet it's surprising that some people ignore it. Those are the people who accept the words 'treasure-hunting' at face value. They buy a metal detector with visions dancing in their minds of massive sea chests with salt-tarnished locks, doubloons, silver goblets, milady's pearl necklaces and the like. They are aware, in a vague kind of way, that the function of a detector is to help to locate non-ferrous articles that are out of sight. They scuttle off to the nearest area of open land, then dart around jabbing the detector head any-old-where. They give the impression that they believe the head is some sophisticated, very powerful magnet that actually sucks valuables up through the soil or mud, to cling to its underside.

Naturally, they find nothing or very little. Within a week, the detector is gathering dust in the loft or hall cupboard, or sold, and the thwarted buccaneer is galloping towards some new hobby.

I have given this over-ripe example to make a point: this is not a pastime for bustling, get-rich-quick fortune-hunters. Such people lack the patience to do it properly; they will not find enough worth-while articles to sustain their first burst of enthusiasm. Those who endure, and therefore make up the great majority of treasure-hunters, are the more thoughtful types who have the qualities that are needed: self-control, application and common sense. People who deliberately loot archaeological sites, or wittingly trespass where they shouldn't, are not treasure-hunters. They are vandals who have got their hands on metal detectors to make their sleazy and anti-social tasks easier. (The image of treasure-hunters in Britain is one of the topics in Chapter 12, which looks into their relations with the country's archaeologists.) Those who already enjoy the hobby, or have commercial interests in seeing it properly conducted, have no truck with characters who have the ambitions and ethical standards of a Long John Silver.

I have tugged that one-legged scoundrel briefly on stage to help to explain why the activity is referred to many times in this book as 'prospecting'. This term has a narrow and somewhat old-fashioned ring to it. In many people's minds, prospectors are those bewhiskered, toothless old sourdoughs portrayed by Walter Brennan in so many films about the Californian gold-rush and Klondike days. This is conditioned thinking once more. Roughly speaking, a prospector is a person who explores a region, or works a mine experimentally for whatever valuable extracts it contains. In the sense of exploring for concealed articles, the business of searching with a metal detector is more accurately described as 'prospecting' rather than 'treasure-hunting'. You prospect every time you take out your detector. For reasons I'll explain a little further on, only a segment of this work is treasure-hunting.

The origins of the phrase 'treasure-hunting' are uncertain; perhaps it does go back to the time of Blackbeard, Captain Kidd, Henry Morgan and other cut-throats of the Spanish Main. Whenever it was coined, the phrase was almost certainly imported from the United States. It is still used there, and rightly so: magazines and instructional books on the subject stress the money aspect. They hammer the theme of hoards and the cash, cash, cash that people can make from treasure-hunting. The phrase has stuck. It is now so strongly entrenched in all countries that, despite its somewhat misleading connotations and essentially phoney glamour, it will be with us forever.

In theory, 'treasure-hunting' is a more significant phrase in Britain than in America. As one professional over here put it: 'Britain is unsurpassed; it is the original treasure island.' The civilised history of the British Isles can be measured in thousands of years, rather than hundreds, as is the case in America. Down the centuries, Romans, Danes, Normans, Anglo-Saxons have all surged throughout the sceptred isle, leaving their imprints by way of culture, language and material possessions.

Consider the latter. It is estimated the Romans brought at least ten times as many valuables and artifacts to Britain as they took away. Hence the profusion of Roman coins, weapons and pottery that a shallowly-thrusting plough blade or even a stubbing toecap can unearth from sites all over the country. The ubiquity of such material brings unceasing anguish and heartburn to archaeologists and puts fierce pressures on treasure-hunters (another point to be covered in Chapter 12).

There is a repeated pattern that weaves itself as people advance in the hobby from newcomer, through apprenticeship to become experienced old hands. What might have begun as vague treasure-hunting (inflamed by those throbbing thoughts about pieces of eight and king's ransoms) ends up as precise prospecting.

To satisfy his curiosity and make his experience as rounded as possible, the searcher should work all manner of sites: houses, footpaths, copses, tidal rivers, ponds, commons and beaches. At first, locating even corroded pennies and horseshoes will be an adventure. The worth-while pieces will start coming and a collection of finds will keep on building—rings, brooches, clay pipes, bottles, buckles, coins old and new, merchants' weights, daggers, shells, buttons and what-have-you.

Then a change in attitude and ambition will very likely take hold of the person with the detector: his interests will focus more and more on specific items he fancies above others: regimental buttons, ammunition, Victorian pot lids, ginger-beer bottles, ancient coins of certain realms. He will feel compelled to go most frequently to those places where he knows he can locate his favourite articles. Treasure, meaning immense riches, becomes secondary.

In this way, treasure-hunting fuelled the host of collecting crazes that sprouted up everywhere around the start of the 1970s. More people were moving about with detectors; a vast array of objects became available for private and public sale, providing the materials for the flurry of collecting booms. These manias ebb and flow. At one given time, clay pipes can be the raging fancy (in part, perhaps, because some of them are startlingly pornographic); then shell casings; then buckles; then handsome cobalt-blue poison bottles.

The instinct for treasure-hunting is universal. It appeals equally to lords who own entire blocks of Mayfair and ditch-diggers whose finances force them to buy the lowest-priced detectors on hire-purchase terms. And why not? The activity tweaks at least two fundamental nerves: the acquisitive nature of man and (this is especially relevant in Britain) his gambling instinct. Everybody delights in finding something. The first item a person finds with his detector will make him quiver with pleasure, even if it's a broken clay pipe, a crushed and tarnished button or a headless toy soldier.

To select and prowl around a new site is something of a gamble, but not a great one if your research is sound enough. When you have learned why certain articles could be in one definite place, the odds are good that they will be there. Of course, if you actually go looking for hoards, you are truly riding your luck.

Your research can be exhaustive and seemingly impeccable. Yet the supposed facts may only be that treacherous old girl, Dame Rumour, tricked out in plausible disguise.

Records are invaluable but a mere error of transcription can be repeated in many references and so throw you off the scent. If you are wise, you cross-check references as much as possible. But, be warned, twenty of them could agree and still be fallible. All of them could omit the most important fact of all: that the hoard you are seeking was stumbled across a hundred years ago by some cagey person who kept the news of his find to himself. It could have been reburied only a few hundred yards away from the original hiding-place, which does have clues leading to it, but this is of no comfort. Despite these hazards, and the possibility of dozens of hoard-searching forays bringing nothing but sweat and frustration, thousands of people who read this book can locate legitimate hoards; they are there to be found.

I state now, and will repeat emphatically several times in the book, that almost all the articles that have ever been lost or hidden throughout Britain's long and vivid history are still waiting to be discovered. Millions upon millions of them. They are your treasure-hunting targets. They lie in wait for people keen, skilled and enterprising enough to find them. Many of these items can be found in places you can reach without breaking the laws of trespass or sabotaging the nation's heritage. This book will tell you where and how to discover them. You must be fully prepared. The first things to learn about are the nature and uses of treasure-hunting equipment.

So let's begin.

CHAPTER 2

EQUIPMENT, CODE OF CONDUCT AND TREASURE TROVE

EQUIPPING yourself for amateur treasure-hunting is not an open-ended business, as it is with some hobbies. Just as the music-lover can satisfy his desire for the company of music through buying the tiniest transistor, the amateur treasure-hunter in his most primitive guise, can get by with nothing more than a trowel, screwdriver and a pair of educated eyes.

You'll note that we said 'get by' with this modest equipment. Without a detector and some of the other basic tools you'll limit yourself to a tiny fraction of the finds that are waiting to be discovered. This would be like taking a job content to remain the mail boy or office junior all your working life. If you buy, and/or make yourself, all the items we list in this chapter the outlay will not be enormous. The only predictable recurring expenses are replacing spent batteries for the detector and fuel costs for your car or motorcycle travelling to sites. Of all the hundreds of hobbies you can pursue, only coarse fishing with worm bait is as inexpensive for the constant enthusiast.

Let's consider the things you'll need for effective treasure-hunting.

DETECTORS

One of Britain's most seasoned treasure-hunters reckons your chances of success leap at least 1,000 per cent when you supplement your eyes with a detector. It was the commercial development of the transistor in 1954 that lifted the hobby for the public above its earlier level of being little more than a ramble with a purpose. Before then, stalwarts lumbered around with things such as army surplus mine-detectors. These were both brutally heavy and expensive. Came the lightweight detector and the hit-and-miss element of eyes-only treasure-hunting was much reduced. Here was a device light enough for women and children to cart around and inexpensive enough to transform treasure-hunting into a hobby available to anyone.

You can choose at the outset whether to shrink your bank

balance by, say, £15 for a simple beat-frequency instrument or up to £200 and above for a pulse-induction detector or £40 to £100 for an induction balance model. The main electronic principles for detectors, their strengths and weaknesses, are covered in detail in Chapter 3.

It will pay to work out your immediate ambitions as an amateur treasure-hunter before buying your first machine. You may prefer to go out with a detector-owning friend before getting your first instrument. An excursion like this has its merits. You may have thought of taking up treasure-hunting as a very languid and random pastime, then get hooked on its pleasures and rewards on this first outing. If so, you could be pleased you didn't tiptoe into it with the cheapest detector on the market. (Youngsters relying on the generosity of parents and relatives to buy theirs will very likely have little option.) But steady finds can enable the impecunious, before long, to earn the deposit on a more sophisticated type; further finds should contribute towards payments.

My organisation mails out brochures, booklets and guides, on inquiry, containing details of all types available in Britain, their costs and details of hire-purchase schemes. Do study these carefully to help make up your mind. In spelling out the pleasures, the rigours and the productivity of various treasure-hunting sites (rivers, beaches, etc.) this book should also guide you in your choice.

DIGGING TOOLS

A lightweight folding shovel is the most versatile basic instrument: you can use it as spade, pick and trenching tool. It is a must for rivers, where you'll reap your steadiest rewards, and beaches. These shovels can be had from hardware stores, sporting goods stores or army surplus suppliers. Most of today's detectors can pinpoint your finds; you won't need to perform like a grave-digger on piece-rates to prise them from the ground.

When searching in some terrain—such as fields and rubbish-tips—you'll need the weight and force of the shovel. Otherwise, you can accomplish a great deal of your digging with a small, sharp-edged trowel. We stock digging trowels especially designed for treasure-hunters; they slice cleanly into any kind of soil, ensuring that you disturb the smallest possible amount of turf and earth. For any digging, you must follow the basic golf-club rule: replace all divots. You should aim to make the ground look as unscathed and undisturbed at the end of your day's work as it was when you came to the spot.

SIEVES

Treasure-hunters work on shale and sand beaches; they rummage on the beds of rivers, creeks, lakes and ponds. For dry-land work, you could make do with a garden sieve that has a quarter-inch mesh. You can place this garden sieve within the circle of a car tyre's inflated inner tube, lash it into position with plastic-coated twine and attach an anchor weight. To meet any emergency you'll need repair kit and pump.

But it's better to make your own non-metallic sieve. Even a fumbling carpenter can do this in about two hours. Obtain a sheet of rigid plastic or nylon mesh about eighteen inches square with quarter-inch holes (Fig. 1). Then make a glued or jointed frame about three inches deep and eighteen inches square. Drill holes through the frame at quarter-inch intervals and lash the mesh to the frame with nylon line.

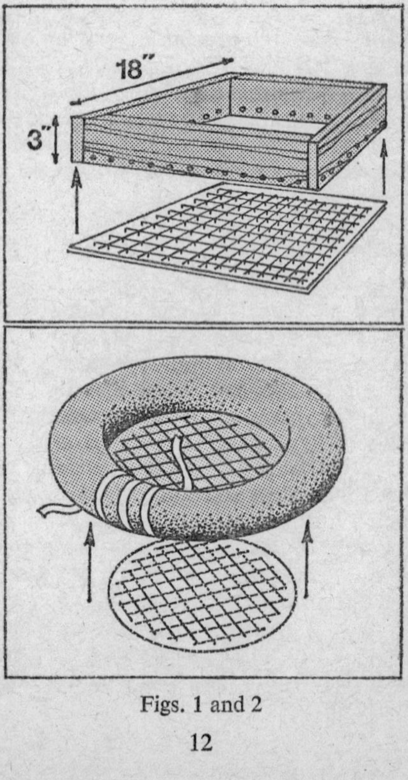

Figs. 1 and 2

For shallow work in water, you should have a floating sieve. As an alternative to the sieve mentioned above, you can place a circular sheet of quarter-inch nylon mesh with thickly taped edges inside the circle of a car inner tube (Fig. 2). Tie it securely to the tube with four or six pieces of plastic-coated wire. The sieve must be anchored with a length of weighted rope to stop it floating away. Plastic mesh is the best material because you can use your detector at the same time to avoid delaying your work. This way, you can search a greater area, and find more articles, in the day's session.

LINES AND PINS

To cover any kind of search area completely and systematically, you must have two sets of lines and pins, the lines about twenty-five yards long. Strong twine or plastic clothes-line will do for the lines, sturdy non-metal knitting-needles or tapered wooden pegs for the pins. The idea is to lay down the lines parallel about a yard apart to form a corridor you can move along slowly and carefully. You'll know exactly where you started and where you're going. Without the lines, even when you are very experienced, your searching would be haphazard and incomplete; you'll never outgrow the need to have them with you constantly.

RIVER RAKES

Do-it-yourself equipment (Fig. 3). The head of the rake is made from two pieces of three-quarter-inch plywood, measuring eighteen inches by six inches. First drill holes down the middle of one piece of ply at quarter-inch centres and hammer some three or four dozen nails—six-inch are the best length—right through. Place the second piece of ply over the flush nailheads and secure the two pieces together with four coach bolts about three inches long. The top piece stops the nail heads from working back into the wood from the pressure of the rake probing the river or pond bed. The five-foot-long handle can be either metal or one-and-a-half-inch hardwood. We recommend the wood; you should try to keep down the number of metal objects that you take detecting with you.

Place the lower end of the handle down the centre of the rake head and drill three holes through the handle and both pieces of plywood. Then secure the handle with four-inch bolts.

Finally, attach a piece of wire-mesh behind the first line of nails (to trap small objects) or weave it back and forth between the lines of nails. Secure its ends with twists of fine, strong wire. The

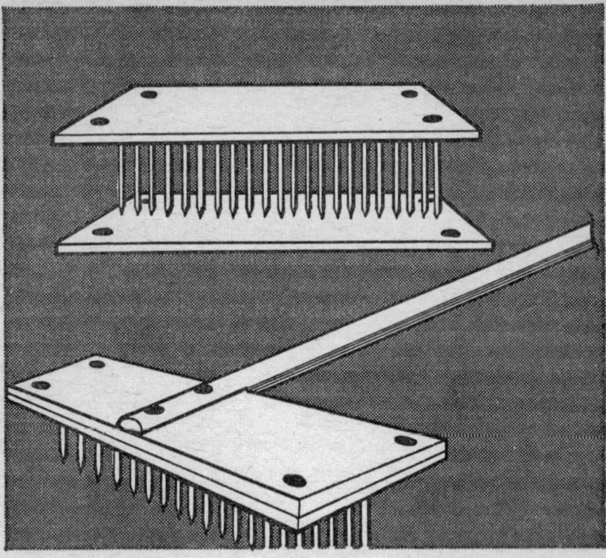

Fig. 3

rake's basic job is to scrape objects from river and pond beds; you can also use it to clear surface nails and metal from river foreshores. You'll need it to probe the depth and check the footing of river bottoms and perhaps smooth out areas of shale or sand you have disturbed.

GLASS-BOTTOMED BUCKETS

These are useful for working in water (Fig. 4). Take a circular waste-paper bin and cut out the bottom, leaving a one-inch ledge. Place on the upper side of the ledge a circular piece of glass that has been cut slightly smaller than the diameter of the bottom of the bin. Now hammer a dozen one-inch nails through the bin sides just above and parallel to the glass to hold it in place. Smear mastic, or some other binding agent, to the ledge both above and below the glass to seal the bucket against leaks. Make a strap from a three-foot-long piece of leather or canvas webbing so that the bucket can hang from your neck as you work.

SCREWDRIVERS

A screwdriver helps when you are treasure-hunting on commons, grass-covered terrain, footpaths and even wet sand. When your

Fig. 4

detector registers and you've pinpointed the buried object, thrust the screwdriver's slim shaft into the ground until you strike something solid. That will indicate how far down the object is and can help to identify it as coin, ring, bottle-top, horseshoe or whatever.

On rough ground, in secluded places where scarring the ground doesn't matter, you can use the screwdriver to prise out the object if it is buried only a few inches down. The most practical screwdriver would be one about ten inches long and with a shaft three-sixteenths of an inch in diameter. For clay, baked or hard-packed soil, a screwdriver about eight inches long with a shaft three-eighths of an inch wide may suit you better. When you've acquired the knack of using one, you can employ it on commons as well; it cuts the turf and flicks out the object very neatly, leaving you a very small divot to push back into place.

CONTAINERS

Always take a small supply of polythene bags of varying sizes on your expeditions. If you're working a river site, one large bag will take the nails and other debris you either rake or skim with a magnet from the surface of the river bed before starting the search proper. You can dispose of the rubbish in a litter-bin or some place where it won't make a litter nuisance; perhaps even carry it home to drop into your garbage can there. A smaller bag lining a pocket or clipped to your belt will hold whatever valuables you find.

When working rivers and beaches, you can place a plastic bag over the detector unit to avoid any dampness reaching it, although most detectors are waterproofed. If thunderheads hang low, you should have a large polythene bag for the whole detector in case a downpour catches you well away from shelter.

For your finds, you'll need something capacious and sturdy, like a converted cricket bag, a naval duffle bag or canvas holdall. Line it with polythene to keep the inside free from dust, mud, leaf-mould, moss, tar or rust marks. The most practical container is a soft canvas sack, like a postman's mail sack to sling over your shoulder. It saves your having to stoop to place either rubbish or finds in ground-level containers. The sack should have large pockets for junk, smaller ones for valuables. You could even make do with a shoulder-strap handbag belonging to a female relative or friend for finds only, but make sure the bag is out of favour and willingly discarded.

CLOTHING

Experienced, dedicated treasure-hunters do much of their searching in raw winter months or the cool dawn hours of summer: beachcombing is done almost exclusively at these times. Thick protective clothing is vital. Wear at least two pullovers, take a pac-a-mac or quilted anorak, sou'wester or balaclava or skiing cap, waders or wellingtons, old gloves and your oldest trousers. You may need a jacket, mackintosh or jerkin for messy work such as industrial river searching.

We don't expect even the most elegant gentleman treasure-hunters, with valet carrying the detector at the high port, to saunter out in two-tone moccasins or patent leather pumps. Something solid, durable and waterproof like hiking boots are probably best, for men and women. Stout work-boots or old snow-boots are also good. It is common sense to wear some kind of headgear for long periods of digging under a scorching sun—

such conditions do prevail occasionally in Britain. When working near water in hot weather, you could be thankful you packed an aerosol can of insect repellent.

ORDNANCE SURVEY MAPS

The original ordnance survey maps of 1874 were drawn two generations before the first primitive metal detector was knocked together in this country. Yet they could have been charted with detector-using treasure-hunters in mind. These maps, which cover every inch of Britain, are so detailed and accurate they are a must, both to use direct on location and to supplement your areas of research. They show you the location of houses, farm structures, wells, roads, footpaths, woods, fences, hedgerows, high and low terrain and a great many forbidden archaeological sites. Earlier ones mark out sites of army camps and bivouac areas, valuable information for the growing numbers of people collecting shell-casings and other ammunition.

Later maps are invaluable for learning of the changes the last century has wrought: re-routed or vanished footpaths, roads and hedgerows, demolished buildings, filled-in wells, parks and commons reduced or enlarged, removed from the public domain or vanished to become residential suburbs and housing estates. For hoard seeking, they also reveal forests that have been cleared.

You can buy them at stationers around the country or from Her Majesty's Stationery Office. They can be obtained on order or you can locate them in your local council library, photostat those available for a fee and arrange to have photostats sent from libraries in regions you wish to search. All in all, these maps are second only to detectors for educated prospecting.

MISCELLANEOUS

Always pack some food. For a brief expedition in moderate weather, a chocolate bar, a sandwich or some water biscuits may be enough; you know your own appetite. You'll need something to replace energy burned up in walking, digging and perhaps climbing.

No matter where you search, it usually won't be within easy reach of the Savoy Grill or even Fred's Transport Caff. On a November beach, the refreshment kiosk's shutters will be creaking or banging in the icy wind. The compleat winter treasure-hunter will have two vacuum-flasks, one filled with hot soup, the other with tea, coffee or hot chocolate. A hip-flask may help to fend off frostbite, and revive sagging limbs, but some purists contend that sipping liquor shoots large holes in your concentration, so they shy clear of it on the job.

For mucky work on riversides and beaches, have cotton waste or rags to clean off hands, boots, finds, screwdriver and knife. A sharp, strong-bladed knife, such as a scout knife, is most useful anywhere but will get most service in cutting out plugs of earth in fields and commons. River and beach work can be very cold. A pair of tightly fitting surgical gloves, well secured at the wrists, will keep your hands somewhat warmer.

To be really thorough, take a notebook and pencils with you. When doing a reconnaissance of a beach, it helps to list the most likely search areas (piers, groynes, bathing-huts, mooring points, etc.). When you go back to work the deserted beach in autumn and winter, you may cover three of these areas in a solid day of toil. Tick them off your list and head straight for the fourth the next day or visit.

Sizing up promising spots from a river bridge, you should sketch diagrams of eddy currents, promontories, landmarks along the banks. At the end of a day's search on common, footpath or wooded clearing, you'd be advised to mark down precisely the point where you stop working (by the twisted oak tree, level with the southern end of the yellow-painted park bench). Next day, you can resume without wasting time pondering whether you missed some ground or are duplicating some of your previously-worked terrain.

MAGNETS

Have a four-inch horseshoe magnet to pull up the nails and other non-ferrous materials from the surface of mud or shingle during river work. It's faster, cleaner and more effective than raking. You can buy a magnet cheaply at a hardware or radio-repair shop; repairmen get them from the backs of television-set loudspeakers.

FOLDING STOOLS

To prospect woods most thoroughly, and therefore lucratively, you should take a lightweight folding footstool, sturdy enough to bear your weight. This will enable you to stretch your detector head as high as possible along tree-trunks and into their branches.

METHODS FOR CLEANING FINDS

Coins and other articles that have lain in the ground for a long time, perhaps centuries, are often dug up wonderfully preserved. This is because the materials they are made of—copper, silver, bronze, gold, pewter, etc.—had reached a state of equilibrium with the acids and minerals of the soil they lay in for so long. Once you have dug them up, you should clean them as soon as possible. Otherwise they will begin to corrode, or oxidise and tarnish.

For cleaning common date coins and jewellery, electrolysis is a simple, effective and accepted method. The best thing is to purchase a small electrolysis unit. My organisation stocks a unit retailing at around £15·00. It measures nine inches by four inches, plugs into the mains and is so straightforward that children can use it comfortably and safely. The method does not take long to learn.

Each electrolysis unit has detailed written instructions with it. When you switch on the voltage, surface dirt and slime is drawn from the object without any other surface actually touching it. The article is restored to its original condition. This process takes from thirty seconds to two minutes depending on the amount of corrosion and thickness of muck that has to be removed.

Electrolysis is gentle and economical. It is, in fact, the only officially accepted method of cleaning coins, which, technically, cannot be cleaned by old rubbing methods and remain saleable. It does a first-class job with jewellery, pulling away the muck from crevices that you could not hope to reach with the corner of a cleaning cloth. Seasoned treasure-hunters say an electrolysis unit is a must if you coinshoot for old and valuable coins—and who doesn't? You quickly recoup the cost of a unit. If you have no knowledge of rare or common coins, always consult a numismatist or coin dealer before using an electrolysis unit.

MODERN COINS

Some banks will accept grubby coins; others won't; it depends on the attitude of the manager or bank official. To avoid having to hump heavy coins to a bank, or around a number of banks, before having to bring them home again, you should clean them. It's a matter of common courtesy.

LICENCES

Before you can use a metal detector legally, you must have a Pipe Locator's Licence from the Ministry of Posts and Telecommunications. When this book went to press, this licence cost 75p for five years. Your dealer will remind you of this obligation when you buy a machine and advise you how and where to obtain one.

In the strictest sense, the amateur treasure-hunter's Code of Conduct is not an item of equipment. It was drawn up with guidance from the Department of the Environment when treasure-hunting was beginning to become widely popular. Many of its points are covered in other parts of this book, but the Code is worth setting down here in its entirety. If all treasure-hunters follow it rigorously and automatically, they should run no risk of upsetting either Authority or their fellow citizens.

1. Don't interfere with archaeological sites or ancient monuments. Join your local archaeological society if you are interested in ancient history.
2. Don't leave a mess. It is perfectly simple to extract a coin or other small object buried a few inches under the ground without digging a great hole. Use a sharpened trowel or knife to cut a neat circle or three-sided square (don't remove the plug of earth entirely from the ground); extract the object; replace the soil and grass carefully and even *you* will have difficulty in finding the spot again.
3. Help keep Britain tidy—and help yourself. Bottle-tops, silver paper and tin-cans are the last things you should throw away. You could well be digging them up again next year. So do yourself and the community a favour by taking all rusty junk you find to the nearest litter-bin.
4. Don't trespass. Ask permission before venturing on to any private land.
5. Report all unusual historical finds to your local museum and get expert help if you accidentally discover a site of archaeological interest. If you discover any live ammunition, any lethal object such as an unexploded mine, do not touch them. Mark the site clearly and report the find to the police.
6. Learn the treasure-trove laws and report all finds of gold and silver objects to the police. You will be well rewarded if the objects you find are declared treasure-trove.
7. Respect the Country Code. Don't leave gates open when crossing fields and don't damage crops or frighten animals.
8. Never miss an opportunity to show and explain your detector to anyone who asks about it. Be friendly. You could pick up some clues to a good site.
9. If you meet another detector user while out on a hunt, introduce yourself. It is very likely you can teach each other a lot.
10. Finally, remember that when you are out with your detector,

you are an ambassador for the whole amateur treasure-hunting fraternity. Don't give us a bad name.

I should elaborate on the subject of treasure-trove. Any person who finds objects of gold or silver (including coins, plate and bullion) in the soil or in buildings must report the find *immediately* to the local coroner; he does this through the police or a local museum, or can write to the Director, British Museum, London, W.C.1. The finder gets a receipt and the authority he has informed gets in touch with the local coroner.

The coroner later holds an inquest to determine whether the find is treasure-trove. He decides (if the original owner cannot be traced) whether that owner had hidden the object(s) intending to retrieve them later on; he also decides who can be legally classified as the owner of the find. If the coroner decides that the owner had planned to recover the property, it is almost certain to be declared treasure-trove.

In this case, the find becomes the property of the Crown. Should no museum insist on having it, the stuff is returned to the finder; he can dispose of it as he sees fit. For instance, he can ask the British Museum to sell it for him at the best price it can obtain. Should a museum want to keep all or part of the find, the finder receives the full market value of whatever is kept.

The important thing is that the find, or reward from a museum, goes to the individual or group who found the gold and silver objects and not to the owner or occupier of the land where they were found. The coroner could state that a group should share the find but this ruling can be disregarded if the members had decided beforehand that the actual person locating any find should keep the whole lot (provided the coroner rules in favour of the finders).

CHAPTER 3

PRINCIPLES AND CHARACTERISTICS OF METAL DETECTORS

NOT very long ago, as recently as 1968, the business of buying a British-made metal detector was rather like dining at a soup kitchen. The entire menu was very basic, very simple, modestly sustaining and offered virtually no choice. There were just two or three home-made beat-frequency machines around. They were quaintly primitive by today's standards of sophistication and performance and, almost certainly, had been assembled in somebody's living-room or attic workshop or on a small assembly-line.

Certainly, a small cluster of American-made machines were on the market, mostly beat-frequency (BFO) and induction balance (IB) machines. They were, and are, skilfully wrought and reliable pieces of equipment but they were, not surprisingly, dauntingly expensive for most embryo treasure-hunters.

Nowadays, the would-be treasure-hunter has a battalion of machines to choose from, both British and foreign. In its modest way, the metal-detector industry has been one of the country's success stories of the early 1970s. At regular, and shortening intervals, a new or updated British metal detector appears on the market. As a result, the information guides that my firm mails out in answer to thousands of inquiries have grown steadily bulkier and bulkier.

These increasingly-sophisticated machines have shaved down the element of luck in treasure-hunting; they have become more practical implements for a widening range of people. Miniaturised electronic circuits, higher quality casings and design features, items like rechargeable batteries all make the machines lighter, more manoeuvrable, more efficient and more economical to run.

They locate buried items more precisely; their signals probe more deeply through soil, sand, mud and salt and fresh water.

All these refinements have opened up the pastime of treasure-hunting. Children as young as eleven or twelve years of age can quickly master the techniques of using the simpler machines, such as BFOs. Elderly people, whose reduced physical strength would have precluded them from lugging around the earlier and heavier detectors, can use the new, lighter ones for long periods. Many persons, of any age, who are hard of hearing, can discern changes in the detectors' audible signals. So they should not be deterred from taking up the hobby.

Proliferating advances in technology during the past two generations have brought metal detectors within the range of most people's pockets and limited mechanical knowledge. As I stated before, the commercial application of the transistor has brought the most dynamic changes. Yet, oddly enough, the principle of transistors has been known for decades. There is evidence that the transistor was discovered before the vacuum tube, or valve, which it has superseded in all types of electrical and electronic equipment.

It is no small irony, then, that tubes were the first to be developed and put into widespread commercial use. Transistors are so much more useful: they are considerably smaller; they drain far less power; their circuits are more durable than tubes with their well-known fragility—elements that burn out, glass coverings that shatter readily.

For years and years, the transistor was a strangely-ignored idea; so was the metal detector; one mystery obviously flows from the other. It is startling to learn that the first metal detector came along in the early 1840s, when a German physicist, Heinrich William Dove, discovered the induction-balance principle. Progress was sluggish, despite the calibre of some researchers. For instance, Alexander Graham Bell, immortalised and often cursed for having invented the telephone, devised a detector which police used in 1880 to try to locate the bullet that killed U.S. President Garfield.

Development crept forward, almost furtively, until the early 1930s when the first commercial metal detectors appeared in the United States. And massive and cumbersome beasts they were. Feverish research that swept through the entire field of electronics during World War II produced the military mine-detector. When the neglected transistor was finally harnessed commercially in the 1950s, the modern metal detector evolved.

Moving on two decades, we focus on the prospective treasure-hunter of today. He (or she) needs a metal detector to search with any degree of efficiency, and, therefore, success. How does the newcomer go about selecting the most suitable detector?

The first practical step is to obtain literature on all the available detectors. There is no shortage and it is all readily available. Anyone interested in taking up the hobby can, for instance, write or phone to my Biggin Hill headquarters, or my agents throughout Britain. A complete set of brochures, and information guides will be on its way by return of post. When the set arrives, pore through each item carefully to work out the detector that will suit you best. The literature contains all the information you will need, including prices, so there is no need to quote any prices here.

There are many factors to consider in buying a machine. Price, of course, may be paramount. You may feel technically equipped and ambitious enough to begin with an induction-balance or pulse-induction (PI) machine. However, your bank balance and financial commitments may dictate that you will, for a start, have to be content with a low-cost BFO detector.

If purchase price, either outright or by instalments, is no worry, then other things will guide your choice: where you live (i.e. close to rivers, beaches or promising heathland where county fairs used to be held), or the types of items you would most like to find. In reading this book you will learn of the multitude of items that you can find with a detector. Consider what takes your fancy and where you are most likely to discover your favourite objects. As I suggested in the chapter on equipment, it can be beneficial to accompany a detector-owning friend on one or more searches. The sites visited and the hauls located may crystallise your thoughts on what you want to extract from the hobby.

Other factors to consider are: whether the detector will be for the entire family or the sole property of an individual; whether it is likely to be used regularly or just occasionally and your aptitude in learning new skills. If you already have some mechanical knowledge, absorb information easily and your wallet has a healthy bulge, you may feel disposed to plunge in by purchasing an IB or PI unit.

Let's consider the four basic types of metal detector. They are:

Beat frequency (BFO)
Induction balance (IB)
Pulse induction (PI)
Transmitter/receiver (T/R).

The principles of electronic circuitry are complex. It would need a qualified electronics engineer with the literary talent of Georges Simenon to explain them lucidly and in depth. I certainly don't have such imposing credentials or talent. So, I'll outline the principles merely in outline where I think it will help.

Beat-frequency (BFO) detectors are the simplest in design and the least expensive on the market. They have few electronic components and need little skilled labour to assemble. In the United States, for instance, thousands are sold in assembly kits for buyers to put together in their own home workshops or front-rooms.

Their low cost and ease of operation make them very popular; a high percentage of the detectors used by amateur treasure-hunters in Britain and elsewhere are BFOs. They span a wide price range; the more sophisticated BFO machines can cost almost as much as some induction-balance detectors.

They are likely to command a large share of the market for the foreseeable future. The smaller outlay for BFOs in the lower price range means they are almost inevitably the machines that youngsters will use to launch into treasure-hunting. Many parents and relatives can comfortably afford to buy them as gifts; they are within the financial reach of independent or enterprising teenagers willing to put aside some of their pocket-money or salaries for a metal detector.

There are two other advantages, apart from price, in starting with a BFO machine. First, the person who takes up the hobby experimentally, and finds that it doesn't suit his tastes, stands to lose virtually nothing—even the time he has devoted to searching. His financial outlay has been small; he can recoup part of it by selling the unwanted machine second-hand; he may even earn enough money from finds to wind up with a small profit. Second, an absolute novice, who handles it properly, can begin to use a BFO effectively almost at once. So the first finds should turn up quickly, kindling enthusiasm and confidence.

Now for a few words on how the BFO machine works. Basically, it consists of two oscillators. The first is reference oscillator, tuned to about 100 kHz; this has a fixed frequency. (Incidentally, for the uninitiated, 1 kHz = 1,000 cycles per second.) The second is the free-running oscillator, tuned to 100·5 kHz, with part of its tuning circuit formed by a coil in the search head (the disc at the lower end of the detector shaft). So the tuning of the second oscillator varies.

A metallic object passing close to the wire coil inside the search head changes the inductance of the coil and alters the frequency of the free-running oscillator. To make this small frequency change audible, and send its output through a pair of headphones of a small radio speaker attached to the shaft, the two frequencies must be 'beaten' or mixed together. When they beat together, one oscillator is tuning to the other, producing a low frequency note. So you get a change in the frequency of the audible note, which rises and registers immediately.

For the best sensitivity, the oscillators must be run at as high a frequency as possible. In Britain, the legal upper limit is 100 kHz. Unfortunately, as the frequency rises in BFOs, so do problems if the buried object lies below sand, grass or soil which is damp. The detector's maximum range (up to two feet for a superior BFO in good soil conditions) can cut back an inch, perhaps more, because the dampness between detector and object has absorbed part of the signal. In some cheaper BFO units, the signal note can warble constantly if the head touches grass or undergrowth, either wet or dry.

Wet grass can also produce false responses in the machine; the signal changes when there is no metallic object beneath the search head. A Faraday or electrical shield which some manufacturers put around the coil of higher priced machines can cut out these false signals to some degree, but the metal shield also wastes some of the unit's sensitivity.

The note of some BFO units is prone to drift, because of the simple nature of the circuitry and through such things as changes in temperature. The note will drift as the machine grows hotter in prolonged sunshine or when the treasure-hunter moves from sunlight into shade, and vice versa. Temperature changes make the search coil expand or contract, bringing a change in signal; hence the drift. So the tuning knob has to be repeatedly adjusted.

The more expensive BFO units use a crystal-controlled reference oscillator which is very stable. However, it is usually very difficult to make the free-running oscillator completely drift-free. Induction-balance and transmitter/receiver units, which also have coils, can suffer from drift away from their selected tuning and have to be adjusted. Because they use radio frequencies (hence the need for a licence), these three systems are sensitive to small objects but give strong signals to such things as coke and silver paper in the ground.

Through the limitations of the system, BFO machines have not improved greatly in range and general performance since the late 1960s. However, better transistors, the introduction of miniature electronic circuits, a wider range of plastic components and good quality headsets have made them lighter and easier to use. Their battery drain is low, another asset for the youngster or the novice merely giving the hobby a trial.

As mentioned before, the beginner can become proficient with a BFO machine very quickly; many units have only one tuning control knob to become acquainted with. These units are sensitive over almost their entire search loop area; they pinpoint objects splendidly. Their decisive and almost instantaneous response to a metal object is another bonus for beginners. The detector user can sweep the search loop fairly rapidly without fear of missing something through any lag in the machine's response.

The sensible treasure-hunter will reap steady finds with a good BFO unit. It is effective on river sites. However, the searcher should be wary at the water's edge unless he has a superior model with a shielded coil. Beachcombers can use them with success. When using some cheaper models, the treasure-hunter is advised to work mainly in dry sand on the upper slopes of beaches beyond the high tide line. The better BFO models work effectively on wet sand.

With some superior machines the user can at times discriminate between buried ferrous and non-ferrous objects. The machine should be tuned to its most sensitive condition so that the treasure-hunter can just hear the oscillator ticking. If the machine passes over a coin or ring, the note will rise to a high pitch. A piece of iron will reverse the signal and the ticking will stop. A word of warning: sometimes this method will discriminate iron; sometimes it won't. Much depends on the size and shape of the hidden ferrous object.

Induction-balance (IB) detectors are much more complex and sophisticated than all but the most expensive BFO machines. Under good soil conditions, the IB can detect a tenpence piece up to ten inches. Detection range depends on the nature of the ground. The range is cut back a little in wet soil and on areas of beaches saturated with wet sand. In soaked grass, the note of some models can warble and give spurious signals as blades of grass brush the underside of the coil. As with BFO machines, a Faraday or electrical shield sharply reduces this nuisance. The design and quality workmanship of most IB units available in Britain are of such high standard that these problems have been largely eliminated.

The more expensive IB units have discriminating circuits that allow the user to distinguish between magnetic objects, such as nails or mineralised soil, and non-ferrous metals. The makers of one high-class machine state that it will detect small objects such as coins, rings, etc. down to nine to twelve inches; larger items such as coin clusters, tools, buckles and knives to twenty-four to thirty-six inches and (glory day!) a treasure chest to a depth of five feet.

IB detectors can be adapted to every type of terrain, so they are excellent all-round devices. They are less sensitive than other types to pins and pieces of wire but will detect large ferrous objects and irregularly-shaped pieces of tin. If you pass an IB unit over a nail, the response of the machine tends to go the opposite way than it does for a non-ferrous object. Practice will train you to discern the differences.

Like BFOs, induction-balance machines can be prone to drift through temperature changes. Let's suppose the treasure-hunter tunes the meter to read zero; this is almost routine. When the search-coil assembly of the machine heats up, through exposure to sunlight, for instance, drift can move the pointer on the meter to give a positive reading. Cooling the search head—by going from sunlight into shade—can move the pointer to negative.

This behaviour is not universal in IB units. Some have virtually eliminated drift. In others, drift can occur for completely opposite reasons to those outlined above: increased temperature brings negative drift and lower temperature a positive drift. Inquire about these characteristics when buying an IB machine. When you are actually searching, it is sound advice to stop every ten minutes to check that you still have the machine tuned to zero condition.

Certainly check immediately you reach a site if you have carried your detector there inside a car with the heater going; this will most likely be the case for six or eight months of the year in Britain. When you take the detector from the car, the search-coil assembly will cool down (this will be very fast in winter months), promoting, most likely, a negative drift. This has to be adjusted before your searching begins.

Drift can occur in some lower-priced machines (both IB and BFO) when voltage drops as the batteries run down. Higher quality machines should have voltage stabilisation circuits built in to keep the voltage constant in critical parts of the circuit.

How do IB machines work? The induction-balance principle normally uses two (sometimes three) coils which are stacked above one another in the search head. These are aligned so that the transmit coil and receive coil are in a delicate state of balance. Alignment and setting up of the coils is critical and usually done through an oscilloscope. Skilled labour and meticulous craftsmanship are needed for manufacture, so the retail price cannot be inexpensive.

When the detector head passes over metal, the electrical balance of the coils is disturbed; the difference is amplified into a signal in the receiver. The signal is either registered on a meter or heard through headphones. The sound comes through as an unmistakable 'ping' when the head is right above the hidden object. So pinpointing abilities of the IB are top-class. In these machines, the line of maximum sensitivity is usually offset from the centre of the search head. Some models have a small spot on the top of the head, marking the most sensitive area.

There is a vital point to remember when using a standard IB unit. The search head is ten inches across, but the sensitive central area of the search loop only three to six inches wide. When prospecting, don't scan the search head from side to side, then advance it the full diameter of the head to make your next sweep. If you do, the insensitive area of the head will naturally miss a lot—it can be up to 70 per cent—of the buried objects in the vicinity. If you are working backwards, as is recommended, bring the head towards you only four or five inches at most to make a new side-to-side sweep. This will obviously slow down your searching rate, but it will minimise the risk of passing over any buried objects. Work out for yourself which you prefer: speed or a greater haul of finds.

This drawback is much reduced in the widescan IB machines that were introduced to the British market early in 1974. In these units, the search head is smaller than the standard model: eight inches compared to ten. The search and receive coils are the shape of a rugby ball with the pointed ends running north and south from the detector user. Therefore, when the detector head is swept across the search area, all parts of the head are sensitive.

In one way, this is an asset, but pinpointing abilities, which are an attractive feature of the standard IB unit, are looser in the widescan model. With a much greater sensitive area in the detector head, the machine obviously registers more signals but you do lose time because of vaguer pinpointing. It's the old adage of swings and roundabouts. To compensate, you should search in 'grid' style. This means that, when the treasure-hunter hears a signal while sweeping the search head side to side, he should locate the maximum signal. Then he should move the head backwards and forwards to find the strongest signal that way. Where the two maximums intersect, the object in the ground is under the centre of the loop. Using this technique will save a lot of time and aggravation.

Pulse induction (PI) is the newest and most complex method employed in metal detectors. Discovered many years ago, it has been used widely and most successfully for airborne mineral detecting since 1965; for the first time it allowed treasure-hunters to locate ore bodies beneath salt-marshes. The adoption of the principle for metal detectors was largely the work of a young British engineer, Eric Foster. This was an achievement to stir patriots because virtually all previous research and development into detectors had been accomplished in the United States and exported from there.

Eric Foster spent four years (1966 to 1970) working on pulse induction at the noted Archaeological Research Laboratory at Oxford University, experimenting to find a means of producing a PI metal detector that would complement the proton magnetometer in archaeological survey work. He left to form his own company and spent another twelve months developing a practical detector to sell at a reasonable price.

Pulse induction is a radical change from other principles; it is the only type of metal detector that does not use a high-frequency oscillator. So it is extremely stable; it suffers no problem of drift. The PI detector uses only one coil for both transmitter and receiver. The transmitter acts like a conventional electric switch, turning the current in the coil on and off. The coil then acts like an electromagnet. Each time the current is cut off, electrical currents are induced in near-by metal objects.

These signals persist in the metal for a short time after the transmitter has switched off; the signals can be sensed with the same coil. The receiver is switched on only when the transmitter has turned off; this happens more than forty times a second.

The coil does not have tuned currents, so the machine's performance and stability are not affected by the electrical capacity of undergrowth or ground (wet or dry) or sand saturated with sea water, as you will find between the low- and high-tide lines of a beach.

Because it uses no high-frequency signals, the PI unit does not detect poorly conductive objects, such as silver paper or coke. Its aloofness to silver paper and wet conditions make it the top machine for beachcombing. The PI machine can be immersed in salt water and up to eighteen inches in fresh water to search rivers, ponds and lakes. Specially-designed pulse-induction detectors are widely used for underwater searching. Many subaqua clubs and marine archaeologists use them, either for commissioned retrieval work in water or to locate, then explore, shipwrecks. Commercial enterprises like major oil companies employ them in their undersea exploration, construction and maintenance work. Tests have shown that these underwater units can work unimpaired two hundred feet below the surface, which is the maximum safe depth for divers.

One of the standard PI machines, the C400, has fourteen transistors and nine micro-circuits. This intricate assembly gives it a significantly greater sensitivity than BFO or IB units. It can detect an old penny or new tenpence pieces to a depth of fifteen inches. Naturally this performance does not come cheaply; a pulse-induction unit is much more expensive than the three other types which use frequency systems.

On some sites, such as an industrial foreshore, it has problems because of its strong sensitivity to ferrous materials. So scrap iron and nails, which you will not want bothering you in these areas, will give a strong signal. This nuisance is more pronounced in the earlier models. With the Auto-Pulse, which is a newer machine than the C400, and fully automatic, the detector gives a different response going over an iron object from that given over a coin or dress ring. A coin produces one sharp signal; a nail two signals like the double-burring sound made by the dial tone of a telephone.

Using an Auto-Pulse, some very experienced and astute treasure-hunters can distinguish the signals given off by bottle-caps and beercan pull rings from coins or dress rings. Tony Hammond, a professional treasure-hunter whose talents are discussed later in this book, can do this. Eric Foster, who developed the pulse-induction principle, cannot. It boils down largely to a matter of practice. By mid-1974, no metal detector had been devised to discriminate between ferrous and non-ferrous objects with 100 per cent consistency but extremely promising research was well under way.

Because they are sensitive over its entire search head, the earliest PI detector models do not pinpoint buried objects as precisely as types using frequency systems. This defect has annoyed some fairly undisciplined treasure-hunters who have felt constrained to dig unnecessarily large holes to locate objects after hearing a signal. The most obvious way to overcome this worry is to use the 'grid' method that I outlined when discussing IB machines earlier in this chapter. Scan side to side, locate the maximum signal; then backwards and forward until you get the strongest signal. Where the lines intersect, you dig and you don't have to produce a slightly smaller replica of the Grand Canyon to find the object.

As well as this technique, the treasure-hunter must use the gain control on PI machines intelligently for best results. There is no need for me to explain these methods here; they are set out explicitly in the manufacturer's literature.

PI machines have a range of loops of differing sizes. If you are working a beach and detect a signal, you will be bothered by sand falling into any large hole that you try to dig in the process of locating the buried object. You could take off the large, standard head and put on a smaller loop—perhaps only two and a half inches across—that you can actually thrust into the hole. Now you have pinpointing.

Some pulse-induction buyers have modified their machines by making their own tiny search coils and adding a changeover switch. Both tasks are quite simple; many people can—and have—done them at home after receiving brief instructions from the manufacturers. The small coil, perhaps only three-quarters of an inch across, is clipped to the detector shaft. If you dig a hole that won't take the large search head, you put on the small one, flip the changeover switch and you are away.

One disadvantage of PI units has been the very heavy current drain from batteries. With BFO machines, the user can often get 200 to 300 hours of use from a tiny PP9 battery, and with IB machines, 40 to 100 hours from a PP6—very economical. The C400 contains seven HP2 batteries (each costing tenpence or more), which have a life of no more than sixteen to twenty hours. Newer pulse-induction machines have rechargeable batteries which last the lifetime of the machine. The treasure-hunter can plug the machine into the recharger overnight and have enough charge for a full day's searching the following day.

Transmitter/receiver (T/R) detectors are almost unknown among treasure-hunters in Britain. They are designed and built, mostly in the United States, for deep detection work and locating mineral deposits. Because their signals probe many feet below the surface, they are also used for such specialised tasks as searching for crashed aircraft whose remains could be either buried under the ground or concealed from the eyes of helicopter-airborne searchers by the canopy of thick foliage. It would be an extravagance to buy an expensive T/R machine for coinshooting and beachcombing. For this searching you would be using, most of the time, only a fraction of the unit's powers. Advances in the technology of the other three types will, I am certain, prevent T/R machines from making any inroads into the treasure-hunting area.

As the photographs in this book show, the BFO, IB and PI machines all look much like skeletal vacuum-cleaners, with a circular head at the end of a slender shaft. The T/R has an entirely different appearance. It consists of two rectangular cases (one upright and the other horizontal) mounted at either end of a connecting-rod. Unless you want a machine for purely gold or mineral prospecting, that is as much as you need to know about them.

In general, no matter what type of machine you are using, if you get a signal you can be confident a metallic object is buried near by. However, now that thousands of people use metal detectors for treasure-hunting, it is worth mentioning a rare problem that I'll call 'phantom presence'. Most searchers will never encounter it; a few may and will be baffled and irritated by it.

Very often, when an object has been buried for a long time, it sets up an electrolytic effect; the soil around the object becomes mineralised to some extent. In very acid soils objects corrode much more rapidly than in neutral soils. Corrosion of a coin in mineralised soil can produce a signal extending for several inches around it. With some metal detectors you will infrequently get a signal from mineralised ground even if the object has corroded almost to nothing or been removed from the ground; hence the 'phantom presence'.

Remember that the metal detector is not a toy, or a magic wand. It is a practical tool that can only be as effective as the person using it. Buying a very powerful and fast sports car does not automatically qualify the owner to compete at Le Mans or in the Indianapolis 500. Knowing how to handle it is what really counts.

The same applies to metal detectors. The mere act of buying, or hiring one, does not put you on the threshold of riches, or even a paltry handful of mundane finds. The formula for success can be divided neatly in half; it is 50 per cent having a good machine and knowing how to use it; 50 per cent employing it on a good site. You must learn, through the ways that are stressed over and over in this book, where these sites are. A steady stream of finds will come only after practice and intensive research.

The first thing to do is to read the manufacturers' instructions very carefully until you know intimately what your machine can achieve and how to operate it under all conditions. Then, before rushing off to a riverside or beach, practise with the machine. And five hours' practice is infinitely better than five minutes' worth.

Don't practise in your back garden because there will be a lot of buried and confusing junk there. Find a piece of open ground or a sand-pit and bury a few test objects there—a variety of coins, some nails, can pull rings, silver paper, all within the known detection range of your machine. Scan the machine over them to learn how it responds; try to distinguish the signals you receive from the different objects. Even if you put a lot of time into these test efforts, bear in mind it could take weeks of actual treasure-hunting before you begin to do really well.

I have devoted several thousand words in this chapter discussing the way metal detectors work and what they can achieve. Like anyone covering this subject, I have done so with some trepidation. Electronics is in the advance guard of the forces that are changing our lives at an ever-increasing rate; they are producing the sense of bewilderment and inability to cope that is known as 'future shock'.

Detectors that were excitingly new and capable when I first became interested in them are now sadly outdated. Those which are talented newcomers to the scene as I write these words could be regarded among research engineers as antiques when you read them. Changes will be largely confined to styles and shapes; no sensational electronic breakthroughs are expected in the near future.

Whatever the changes, please don't blame us; blame man's ingenuity and the frenetic whirlwind of change that is the twentieth century.

CHAPTER 4

HOW TO SEARCH HOUSES AND GARDENS FOR HOARDS AND SMALL FINDS

To launch yourself actively into treasure-hunting, you don't need to add to your petrol bills or to court chilblains fossicking along a winter beach. The apprenticeship can begin in your own living-room, kitchen or attic—preferably all three.

If your home is a mere ten years old and you have been the only inhabitant, naturally you won't be searching for a hidden hoard, unless you suffer from amnesia. Yet, even if your home is new, a thorough work-over with a detector can discover the lost tie-pin, slim brooch, cufflinks, propelling pencil or the Victoria Cross you have mislaid. You'll certainly find a startling number of modern coins. When you've exhausted the treasure-hunting in your apartment or semi-detached, or sixteenth-century mansion, you can take the detector to Aunt Emily's suburban villa in Eastbourne to search for the fish-knife she lost during the power cuts in 1947. Then work some demolished houses. (But more about them later.)

Most of us, however, don't need to stir beyond our own portals to begin seeking old valuables. Such has been the explosion of collecting mania since the mid-1960s that items dating back to 1930 have been formally classified as antiques. The chances of a piece of Victoriana lying beneath your floorboards, or behind the skirting-boards, are higher than you think. Thirteen out of every twenty people in Britain live in houses built either in the Victorian era or when Victorian coins were in circulation. In the seventy-five to one hundred years these dwellings have been standing, they can have housed three, eight, a dozen or even twenty different families. The most careful and brilliantly organised of these families lost a considerable number of items; that's ordinary human nature at work.

In Victorian and Edwardian times, middle-class merchants could afford to run homes staffed by four or six servants and crammed with a wealth of very losable bric-à-brac. Those were the days long before fitted carpets, snug-fitting and tailored matting, sealed rubber tiles and other furnishings that deprive dropped articles of convenient hiding-places.

In those times, housewives or cleaning maids scrubbed and polished floors in all rooms and passageways. The flick of mop or scrubbing-brush could send the half-sovereign, pocket-knife, pendant or whatever skittering down a floorboard crack or under the skirting. Even when the loss was noticed—and a great many were not—the fugitive article would elude a systematic search because it nestled in some dark corner beyond the arc of hissing gas-lamp or fluttering candle. Afterwards, a toecap or piece of moved furniture could have knocked it out of sight to some place where it is still lying.

So use your detector as a kind of indirect vacuum-cleaner from the roof beams of the attic to the cellar floors. As always, work slowly and intently. You'll realise more than ever that older houses have a multitude of niches and recesses for lost and hidden articles: hollow cupboard doors and bedknobs, backs of mirrors and picture-frames, rear sections of fireplaces and inside chimneys, hollowed-out joists, above and behind beams and rafters and also under floorboards and stairways. (Crosses in Fig. 5 show some of the hiding-places inside a house.)

While working all rooms and passageways, pass the detector head carefully along the bottom of the walls. You'll quickly identify the signal of nails and pipework; before long you can discern the bleep of lost coins. Study the areas where floor joists sag and have opened a space beneath the skirting-boards. Work a piece of hooked wire well into the gap, keeping it level with the floorboards, then draw it slowly towards you. Then move along a few inches, thrust the wire in again and repeat the procedure. Work your way around the room, and all others, using this method wherever gaps allow you to do so.

Locks in old houses are worth checking; children have always found them enticing places to stuff tiny coins and trinkets. It is thought that every fifth lock in British homes contains at least one coin. One prospector tells us he scooped twelve Victorian coins, including a rare farthing and half-sovereign, from inside a single lock. To reach the coins, you'll have to unscrew the lock, take off the backplate and turn the lock upside down. Whatever is inside will drop out. We don't recommend that you go lock exploring like this when you're a guest. Even in your own house it would be best to wait until you're redecorating, then fresh paint will cover the marks of your handiwork.

For some reason a lot of people baulk at giving the loft or attic

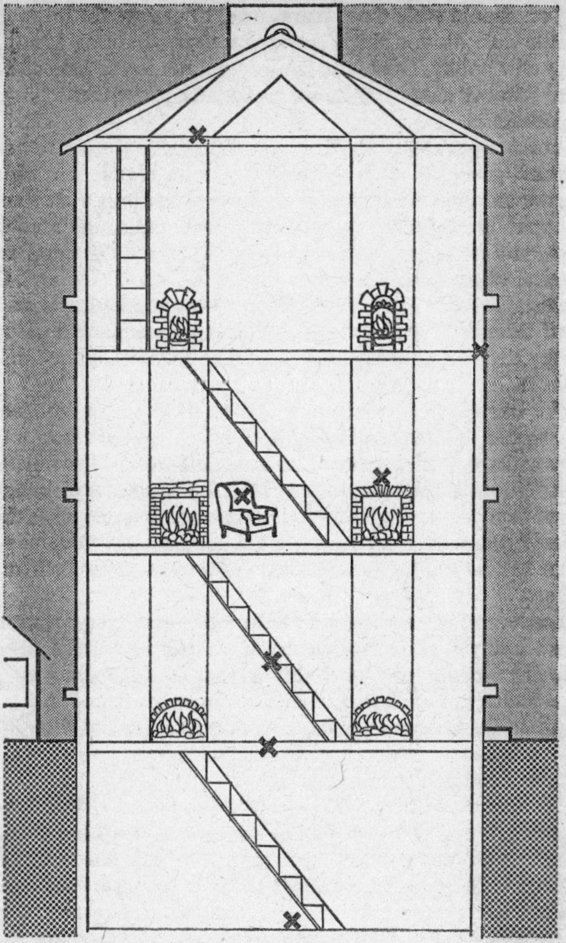

Fig. 5

a really thorough going-over. They envisage themselves threshing around in cobwebs and sneezing amid the dust puffs from disturbed piles of old newspapers, boxlids and grandmother's dressmaking dummy. Or they are afraid the ghost of great-uncle Ebenezer will slither out of a trunk when they fling back the lid.

You should stifle these fears. And, if you are fretful of getting a little dust on you, then perhaps treasure-hunting is not really your true hobby. Dust is nothing compared to the mud and damp sand that will cake your hands and clothing when searching rivers and beaches.

Attics are really the most obvious places in the entire house to explore generally with hopes of coming across the unknown Turner painting, Stradivarius violin or mint-fresh Victorian coins wrapped in a piece of faded velvet. With collecting mania rampant, you could make money from finding old *Gem* and *Magnet* comics, cigarette cards featuring Dr W. G. Grace and Enrico Caruso, Donald McGill postcards or the model Lagonda drop-head limousine that somebody needs to complete a set of old Dinky toys. Of course, this is productive browsing, not treasure-hunting.

For our type of searching, work over the attic as diligently as you would the Bank of England vaults. Move the detector slowly over walls, chimney-breasts and above all beams. The nineteenth-century lodger might have been Jack the Ripper, a pickpocket or a well-bred kleptomaniac. While making routine searches through houses, police have prised numerous compact but valuable hoards from behind hollow bricks, fake walls or unscrewed them from rafters.

Leave nothing unscrutinised. In attics and every other room, make sure you glide the detector head over all fittings and furniture; arm-chairs are inevitable repositories for dropped coins, cigar cutters, snuffboxes, watch charms, silver and gold tooth-picks, etc. Cupboards, armoires, vanity tables and whatnots can have false bottoms and hidden compartments; stairways abound with artfully disguised crannies.

To investigate beneath uncovered floorboards, you'll have to lever them up. This is a simple job, not as difficult and cumbersome as you may predict. You will be looking almost exclusively for coins, the most likely items to have dropped between floorboards. Over years, or perhaps decades of constant washing, water will have seeped down, caking the dust beneath and imprisoning the coins in earthen batter. This you must scrape out carefully with a knife and break open. It will be surprising if you end up with only the largest collection of retrieved hairpins in your neighbourhood; the odds against *not* finding coins are very high indeed. Replacing and securing the floorboards is straightforward work. Be cautious when probing underfloor cavities close to walls; older systems of electrical wiring in all too many British homes are illogical and untidy.

Similarly, skirting-boards mask an astonishing variety of stuff: sovereigns, necklaces, jewellery, medallions, letter openers, rings and so on. Searching them, however, is messy. You should undertake it only if you are planning major alterations in your own home or are searching houses marked for demolition.

When you have sniffed the detector over the last barrel and loose brick in the cellar, take it outside to go over every square inch of the gardens, front and back. Remember that children have romped there, adults strolled, courted and picnicked there. Many generations of rubbish have been tipped and burned in some part of the back garden.

Dig deeply into the site of any old rubbish-tip or incinerator; the coins, rings, mustard pots, sugar bowls that were accidentally tossed out with the garbage or shaken from tablecloths could be waiting for you to dig them up. And make sure you check old brick walls, where children loved to hide things, gazebos, summer-houses, rockeries and around the base of the oak tree, where the children soared high on swings and the whole family played hide-and-seek. The sense of family unity has much diminished in Britain since those Victorian and Edwardian days, but a considerable number of their lost and forgotten possessions are still around.

So before you've had to move beyond your own small, or large, plot of land, your mantel and sideboard should glint with your first finds. Outside your front gate, the simplest buildings to search are those vacant awaiting demolition or actually being pulled down. You won't have to travel far to find them. Since World War II, local authorities all around Britain have torn down hundreds of thousands of old houses, shops, factories and warehouses in slum-clearance and redevelopment schemes. Huge property companies eager to wipe out older buildings to replace them with shopping complexes and massively profitable skyscrapers have speeded up the rate of demolition. So, too, have construction companies who have made their fortunes crunching buildings into dust to make way for motorway schemes, underpasses, overpasses and swollen road junctions. Their combined efforts are the mainspring for the anguish and frustration that the nation's archaeologists are suffering. The archaeologists are few, the agents for change many; an uneven battle in which tomorrow's supposed needs are obliterating the surviving traces of yesterday;

so much of our heritage is simply being torn down, excavated and cemented over. I'll go into this dilemma and the tensions it has caused in a later chapter.

To learn of impending demolition, closely scan national and your local newspapers, watch for announcements from your own and neighbouring councils about demolitions. Development plans are often tightly screened while the council is considering applications but ultimately they have to be announced publicly; the borough engineer's department can inform you. To supplement this information, you should move regularly around your neighbourhood, chatting with residents, shopkeepers, publicans. People like bus drivers and conductors will know of demolition works along their routes. You'll pick up a lot of information this way.

Of course, you have to get permission to look over and work building demolition sites. A polite request to the construction company's foreman may set up an arrangement whereby the security guard gives you access to search there briefly under supervision after working hours in summer or at week-ends when the crews are absent. This is a long-shot, but worth trying.

The best period is before or in early stages of the demolition; as it progresses the dwindling building gets less and less safe. Whatever the building, you'll have to fine-toothcomb it with the thoroughness of a Sherlock Holmes. In dockland or slum areas, your concentration must never be less than 100 per cent.

Thousands of slum-dwellers were ignorant of banks or just distrusted them; they clung to the tradition of the breadwinner collecting his wage packet and never putting any of its contents into a bank account. They lived in cheap, sometimes flimsy houses with thin walls, frail and inadequate locks, ill-fitting doors and window-frames. Their houses were easy prey for burglars and small-time thieves, many of whom lived in the same areas. So they felt compelled to hide their valuables as best they could. And sometimes the best methods were very ingenious indeed. Carpenters, chandlers, labourers—men used to working with their hands—fashioned clever, undetectable false bottoms in bureaux and chests of drawers, or scooped out and re-mortared niches in chimneys.

One treasure-hunter, looking over a half-wrecked tenement in London's East End, discovered an entire and perfect false floor in one of the apartments. It would have fooled the shrewdest of burglars. When the astonished, and admiring, treasure-hunter raised the false floor he discovered a number of biscuit tins, empty

The first step. My sales manager, Peter Bettis, describing the features of a BFO machine to a customer.

The thirteen rings on these fingers were all found under the sands of Littlehampton beach in Sussex. Thousands more rings lie under British beaches.

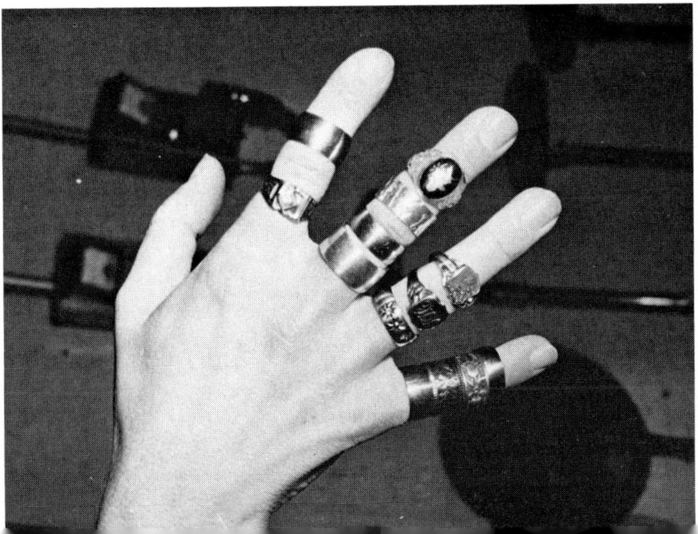

It's a hobby for kids as well. The curious youngsters testing these BFO units can learn within hours how to use them well.

This lass hasn't yet worked out the right angle for her BFO (the head shouldn't tilt) but her concentration is fine.

Shrewd application. This young hoard-seeker stands on a wooden block to stretch the detector head as far as possible into the tree.

Good technique with an IB machine. The user's whole mind is on the job; and the detector head as close to the ground as possible.

John Webb searches a foreshore with an IB between firmly secured line and pins.

John Webb, again, seemingly a little distracted as he sieves in the Thames.

Pubs old and new are fine for treasure hunting. People gather in them in large numbers and, not surprisingly, become careless.

Tony Hammond using one of the best coinshooting machines; a PI unit with rechargeable batteries.

Signals in high frequency detectors can drift. Dedicated amateur, Tony White, here adjusts the tuning of his IB unit.

Tony White uses trowel to dig up find. Note that he has not laid the detector on the ground as he digs.

The PI machine Peter Bettis has here can detect a tenpence piece in wet ground down to fifteen inches.

The BFO seen here has a six-inch detection range, shallow enough for the searcher to use a screwdriver for retrieval.

Large hoards have been found in derelict buildings. No place is too unlikely, even a flimsy wall like this.

More knowhow in searching an old house. Lintels and doorframes must always be carefully checked.

except for pieces of paper showing the sums of money that had been hidden in them.

Some husbands chose not to tell their wives and families of the money they tucked away in such places. In other houses it could still be there; the husband could have died accidentally (electrocuted at work, knocked down by a car) without divulging his hoard.

Demolition has been an expanding fact of life for a generation. But thousands upon thousands of buildings approaching their century will be standing for many years more, perhaps decades. Find out about these occupied houses that may conceal something. Pore over old council records, yellowing newspaper files, ancient almanacs and books of your locality. From all these sources, you'll piece together local history, discovering the sites of many possible hoards.

For example, No. 243 Colwyn Avenue, just a few streets away from where you live, was once the home of a man convicted four times of burglary; he returned there to live between his spells in jail.

Police may not have recovered all his loot; they are the ultimate professionals at this type of work but they are not infallible. The discreet omissions of the landlord or estate agents may have meant that succeeding tenants did not know there was anything to search for in the house.

And what about that many-gabled, spooky old house on the hill not far away? Perhaps there are municipal records and old newspaper cuttings, perhaps surviving neighbours who recall that two eccentric spinsters lived there for forty-five years with a multitude of cats. A perfect set-up for gossip to flow and multiply about the wealth they must have stored away somewhere. Who knows? The rumours might have been not only true, but wildly conservative. After all, rag-pickers who lived in shanties have died leaving behind tens of thousands of pounds and stacks of mildewed securities bound with grubby twine.

If your researches have pointed strongly to a cache of some kind still undiscovered in an occupied dwelling, approach the people who now live there. Explain what you are after; hold nothing back. Should your target be a cedarwood box of sovereigns and nineteenth-century cameos, don't pretend you want to look around for 'a few quids' worth of bits and pieces'. This subterfuge deserves an instant rebuff, and will almost certainly earn you one.

Strike whatever agreement you can with the descendants or

tenants about sharing the spoils. They could demand the lot. If so, it might still be worth while—once and only once—searching to satisfy your curiosity and build up more experience at house searching. This could prove valuable training that could bring you a future bonanza. You will probably find that most people, approached courteously and told the whole story, are likely to be accommodating and generous.

However, to avoid any later acrimony, should you locate something, put your full agreement on paper for both parties to sign. Sudden riches may enliven the heart, but they can also poison the conscience.

Those lucky treasure-hunters who come across a hoard in house, copse, field, anywhere at all, should let the newspapers know. The romance of concealed hoards makes them fascinating news stories. The report will spark off interest locally, perhaps nationally. It could bring you a flood of information about possible hoards elsewhere; if so, the ratio of chaff to wheat will be about fifty to one. Of course, many people will be intrigued enough to begin searching for themselves.

This will hasten the inevitable growth of the hobby—along educated lines, we hope. There could be other people too frail, too uncertain, too scared of the dark to search their own cellars or attics (they may simply be honest enough to admit they don't know the techniques), who will invite you to look over their houses.

In these cases, make sure the householder is not asking you to waste your time following his vague hunch that great-aunt Agatha might have left behind a hidden hoard of sovereigns because she was notoriously mean at giving him pocket-money. He might have been an obnoxious little boy who didn't deserve much. Don't accept such an invitation until you are experienced enough to discern whether such evidence as the householder presents is either promising or worthless. Any worth-while search must be thorough; it will consume a considerable amount of time that you may be better advised to spend coinshooting on your favourite beach.

Houses can certainly pay off for you. But don't be obsessed by them. By all means, work every demolition-marked building you can, but only apply yourself to try an occupied building or dwelling that gives powerful promise of leading somewhere. The complications are many: seeking permission, striking binding agreements, raising floorboards, chipping away and restoring chimney mortar, perhaps breaking open neglected and locked rooms whose keys are missing.

All such effort makes this kind of searching a far more difficult task than, say, river searching. In rivers, the tides are your servants and unfailing delivery-men. They can threaten you but there is no danger they could default on you and dismiss you empty-handed.

CHAPTER 5

DESCRIPTIONS AND LOCATIONS OF HOARDS FOUND IN BRITAIN

IN explaining the reasons for and the ways of searching a house, in the previous chapter, I had a lot to say about hoards. Now I want to elaborate on that topic even more.

Hoards are the absolute cliché of treasure-hunting; when you utter those two words, the unenlightened listener is most likely to assume that hoards are all you crave and search for as a treasure-hunter. This is not so, but hoards are indeed an integral part of the activity. The reason is simple: there are so many of them around. More than four hundred hoards have been discovered in Britain this century. This figure could double, triple or even increase twentyfold before the twentieth century bows out. The hoards already found are only an infinitesimal fraction of the ancient and modern wealth that has been deliberately concealed and remains undiscovered. The number of unfound hoards could run into tens of thousands.

They have been, and will be, found everywhere: in houses, caves, barns, mills, towers, woods, fields, ditches, rivers, trees, cliff-faces and under moats, footpaths, fords and beaches. The finders have happened upon most of them by accident, even when the location and circumstances might have suggested hidden valuables.

For instance, in April 1969, the Reverend Stanley Jones, vicar of the parish church at Wybunbury, near Nantwich, Cheshire, was spring-cleaning the fifteenth-century church tower with his son. They opened a chest that had been lying there undisturbed for many years and found antique silverware valued at more than £26,000. The most valuable item was a lidded Charles II tankard dated 1677 and believed to be worth more than £10,000.

A few years earlier, four-year-old Christopher Forrest was playing in the garden of his home in Tadley, Hants, when he kicked a rusty tin shallowly buried near a hollow pear tree. Inside the tin were a total of twenty guineas and half-guineas from the reign of George III. Unhappily, the local coroner ruled that the tin had most likely been lost and Christopher could not keep its contents as treasure-trove.

Perhaps it is their free-flowing curiosity and sense of adventure that leads so many children to stumble across hoards. Back in 1950, a fourteen-year-old schoolboy, raking earth at the bottom of a ditch which interested him, near his home in Hailsham, scraped up this collection: fourteen sovereigns, eight half-sovereigns, six half-crowns and a Victorian shilling.

Three years later, perhaps spurred by lurid tales of rich misers, a party of thirty Bradford schoolboys decided to spend their school lunch-hour rummaging around a deserted farm cottage at Eccleshill, near Bradford. They levered up the floorboards and hauled out twelve small bags that held 230 sovereigns, 200 half-sovereigns and £20·00 in silver. The value of £900·00 placed on it would have been worth four or five times as much in the 1970s.

In 1968, another fourteen-year-old, John Crossland, for unexplained reasons that might have brought parental wrath, took out some of the floorboards in the bedroom of his home at Tintwistle. Perhaps he escaped a sound spanking by bringing out eleven sovereigns and thirty-four half-sovereigns, dating from 1830 to 1870 and all in mint condition.

Master Crossland made his find a few months after two lads unearthed a pot from a field at Bolton Percy, near York. This contained 1,500 coins known as stylas and all bearing the names of Northumbrian kings and two Archbishops of York. The hoard was thought to have been buried there to conceal it from Viking invaders.

An adult, Mr Malcolm Tricker, also shared the luck which ran so strongly in 1968. He was operating a bulldozer on a building site at Belstead, Suffolk, when he found five gold torcs, or collars, dating from 100 B.C. in the Iron Age. Experts at the British Museum valued his finds and then, the following year, awarded Mr Tricker £45,000 for his discoveries. In 1970, Mr Peter Graham dug up a similar torc in the garden of his home in Holcombe Crescent, Ipswich. His neighbours immediately rushed for their shovels and sent earth flying from garden plots right along the street. No further finds were reported.

Massive road-building and -widening schemes and redevelopment projects, which have become so much a part of contemporary life, have been possibly the most constant sources of revealed hoards. Almost every week, workmen wielding picks, mattocks and shovels, or at the controls of giant earth-moving machines, disturb a cache from its slumbers of hundreds or thousands of years. It is not surprising that growing numbers of these workers have bought metal detectors to prospect before or after their work hours or during their lunch-break.

Here are just a few of their finds:

- Workmen digging drains near Ramsbury, Wiltshire brought up an earthenware pot containing sixty-one coins, including a Charles I crown, sixteen Charles II crowns and forty-four half-crowns.
- In 1962, builders renovating a large detached house at Disley, Cheshire hauled down an old biscuit tin from the rafters. Neatly bundled inside, and bearing a label 'soiled notes', was the sum of just under £1,700 in banknotes, £1,500 in one-pound notes and about £200·00 in ten-shilling notes.
- A mechanical digger driver, in 1964, unearthed a total of 1,108 silver coins (pure silver groats, half-groats, pennies and halfpennies dating from 1450 to 1503) while doing roadworks at Hertford. Some of the coins were in mint condition. They were thought to have been the proceeds of a theft and buried in two pots, one standing on the other.
- A construction manager found traces of a presumed leather purse and more than 1,000 gold coins four feet under the surface of a building site near the gates of Newstead Abbey, Notts., once the home of Lord Byron. With the coins (one dated 1199) were a cross, three rings, a locket and chain, and a brooch.
- In September 1966, two Manchester Corporation workmen, digging a ditch in a meadow at Portfield Farm, Whalley, Lancs., located a gold armlet, gold hair-ornament (called a tress ring), a carpenter's gauge and a number of bronze axe-heads and other tools. Museum experts believed that Bronze Age craftsmen had hoarded them underground between 700 and 800 B.C.
- In 1968, a digger driver struck an old earthenware pot containing 415 silver coins of 1550–1650 on the site for a new school at Sheerness, Isle of Sheppey. The coins bore images of Edward VI, Elizabeth I and Charles I. The British Museum declared the workman could keep the hoard, worth many thousands of pounds.
- In February 1969, two labourers dug up 14,105 medieval silver coins in a heavy lead casket on a building site at Colchester. They were mostly long cross Henry III pennies, dated between 1247 and 1279, plus some Scottish and Irish coins. The coins were later returned to the workmen. It was thought that a thirteenth-century official had buried the coins about 1280, intending to re-

turn them later for renumbering. The theory was put forward that the official secreted them and had been killed before he could recover them. Discovered beneath the foundations of a sixteenth-century building, the hoard was possibly the largest collection of medieval silver coins preserved intact in Britain. In 1902, a similar hoard of 11,000 silver coins had been discovered just a few yards away from this one.

● In February 1972, a bricklayer brought up twenty coins of the reign of Henry I from a trench he had been digging. Experts called to the spot, in Lincoln, located another 730 coins.

● A contractor digging pipes at Warsop, Notts., found 341 Roman silver alloy coins minted between A.D. 315 and 330.

As workmen keep thrusting the blades of manual and mechanical implements into the face of Britain, so these hoards will keep coming to light, possibly at an accelerating rate.

All the finds mentioned above came from the soil; the water also yields its share of booty, some of it extremely valuable.

Some years ago a London house-painter and his eleven-year-old son were strolling along the Thames mud-flats at low tide in the Battersea area (whose attractions for treasure-hunters are described in Chapter 7). The father saw a glint of something in the mud. Bending down, he scooped aside a few handfuls of mud and tugged up a ruby-studded bracelet, a silver cup, candlesticks, an assortment of clocks, a tray and several snuffboxes. Total value: £1,000.

If the spirit of Lord Nelson roves abroad from his resting-place in St Paul's, it may ponder what happened to his effects after his death at Trafalgar. Within a span of five years, parts of two sword scabbards were recovered from different rivers; both were thought to have belonged to Nelson. In September 1968, Mr Fred Besch found part of a diamond-studded gold scabbard in the River Wey, near Farnham. Two large stones had been prised out but 633 small diamonds remained, set in enamel and gold. Then, in November 1973, Mr Roger Coyle, diving in the Thames near the Albert Bridge at Old Windsor, pulled an object from the mud on the river-bed. It turned out to be a diamond-encrusted chape, which is the tip of a sword scabbard. Like the other, it was thought to have been Nelson's. A short while later, a jury at Maidenhead said Mr Coyle could keep the chape, valued at £36,000.

Nelson had certainly not lacked swords; he was presented with five after the Battle of the Nile. In 1803 he made a will leaving a sword, musket and canteen—all gifts from the Sultan of Turkey—to a friend, Alex Davison. How portions of the sword scabbards turned up in rivers well apart is an intriguing mystery; but perhaps most unidentified swords are automatically assumed to have belonged to Nelson until it is proved otherwise.

In the following chapter, I will explain why the treasure-hunter should not search for any hoard unless he has researched so thoroughly that he is following a specific trail towards something he has strong reasons to believe really does exist. Pursuing mere hunches is worse than useless; the odds against coming haphazardly across a hoard are very high indeed.

An example of this deflected luck occurred in November 1971. Two computer lecturers were treasure-hunting in woodland beside the A40 at Aston Rowant, looking for Charles II coins. On their first visit, they found two Roman coins and a brooch. Later they came across a hoard of Anglo-Saxon silver sceattas, thought to have been the life-savings of a peasant of that era.

It was the largest single hoard of its kind ever found in Britain. Each of the coins, about the size of a new halfpenny, was worth between £30·00 and £50·00; the hoard was valued at between £5,000 and £8,000. According to historians, an Anglo-Saxon peasant could have lived for a week on one of those coins.

All the hoards and finds described earlier in this chapter were discovered during the last generation. Below I'll list briefly other hoards, mostly ancient, that have come to light over the past hundred and fifty years. They were found in a variety of locations and some of them had lain dormant for more than a millenium. I think they prove graphically my earlier point about the staggering amount of wealth yet to be retrieved from secret places throughout the country.

Area: Balcombe, Sussex. Found: 23 May 1897. Buried: 1380.

More than 700 gold coins were discovered in an iron container, which looked, strangely enough, like a modern coffee-pot.

Area: Beauworth, Hampshire. Found: 1833. Buried: 1087.

It is thought workmen came across this immense hoard of 12,000 silver coins in a large lead box close by the local manor-house.

Area: Beaumont, Cumberland. Found: 1884. Buried: 1360.

This hoard contained almost 2,000 silver coins and an assortment of groats, half-groats and a few foreign coins.

Area: Blackhills, Aberdeenshire. Found: 1911. Buried: 1320.

This hoard had been placed in a wooden bowl, a most unusual and not very practical container. Its 2,026 pieces were mostly silver pennies, with a few English, Scottish and foreign coins besides.

Area: Borscar, Dumfriesshire. Found: 1900. Buried: 1322.

This had 1,382 pieces, just a little more than one for each year A.D. at the time it was buried. They were silver pennies, English, Scottish and foreign, in an earthenware pot covered with cloth that was surprisingly intact. This suggested it was in clay soil or peat.

Area: Borth, Cardiganshire. Found: 1930. Buried: 1425.

A collection of thirty-one gold nobles that were loose in the ground and said to be in very good condition.

Area: Boyton, Wiltshire. Found: 20 July 1935. Buried: 1324.

The 4,147 coins in this hoard were mostly English and Scottish silver pennies and mules. A few were counterfeit.

Area: Bredgar, Kent. Found: Easter Monday 1940. Buried: 1388.

This hoard of 131 gold nobles, half-nobles and quarters was found in chantry cottages not long before the Battle of Britain brought another kind of excitement to this part of England. Oddly enough, the finder thought they were of no value when he first located them.

Area: Canterbury, Kent. Found: 1901. Buried: 1130.

While the number (358) and kind (gold) of these coins were recorded, the exact finding place was not. But this omission is not as unusual as it may seem at first glance.

Area: Carsphairn, Kirkcudbrightshire. Found: 1913. Buried: 1330.

Consisting mostly of silver pennies (English, Scottish and some foreign coins), this hoard of 2,225 coins was in a pottery jug that had been broken, very likely by ploughing.

Area: Chanctonbury, Sussex. Found: 21 December 1866. Buried: 1066.

A total of 1,720 silver coins buried the year of the Norman Conquest and a splendid Christmas present for its finder exactly 800 years later. They were in a crock that, it was felt, had originally contained many more than the recorded number. Most likely the missing coins—if any such had existed—were sold privately and discreetly.

Area: Chester, Cheshire. Found: 1950. Buried: 970.

Workmen laying cable discovered this historically important hoard. It contained 525 silver coins, mainly Anglo-Saxon pennies and a few foreign items, plus forty bars of silver. The hoard was estimated to have been buried during those two centuries of neglect in Chester between the ravages of Danish and Saxon occupation and the time the Normans revived the city. Archaeologically and architecturally, it is as fascinating as any city in Britain. Because virtually every stone and cornice is of historical significance, it is definitely one place for the responsible treasure-hunter to leave alone.

Area: Cleuchhead, Roxburghshire. Found: 1897. Buried: 1310.

There 138 silver pieces found here were all foreign, called 'sterlings'.

Area: Closeburn, Dumfriesshire. Found: 1844. Buried: 14th century.

An enormous amount (10,000 pieces) of English and Scottish groats and pennies found in a tripod cooking-pot.

Area: Colchester, Essex. Found: 1902. Buried: 1260.

One of the largest hoards ever of medieval coins (10,926) was discovered in a shallow lead box in Colchester High Street. Like Chester, it is another ancient (dating back several centuries B.C.) city. Not a place for treasure-hunting, but rather for visiting the Colchester and Essex Museum to see local exhibits ranging from the Stone Age to the seventeenth century.

Area: Coldingham, Berwickshire. Found: 1853. Buried: 130?

The hoard of 693 silver coins, English and Scottish pennies, had no recorded vessel or burial container.

Area: Congressbury, Somerset. Found: 1828. Buried: 1470.

A total of 138 Henry V and Henry VI gold nobles and silver groats were found, also a single Edward IV gold royal.

Area: Corriemonie, Inverness-shire. Found: 1870. Buried: 14th century.

Appropriate name for a place where a hoard of 570 silver coins was found more than a century ago. The person who buried these silver pennies was very painstaking: they were all packed neatly on edge inside a copper pot that was dug up in the local churchyard. Corriemonie, also spelt Corrimony, is a fascinating spot for antiquarians; it has a neolithic passage-grave inside a chambered cairn.

Area: Guisborough, Yorkshire. Found: 1848. Buried: 1480.

A smallish hoard of 195 silver coins: English and Irish groats, half-groats and pennies.

Area: Halsall, Lancashire. Found: 1920. Buried: 1428.

This hoard consisted of twenty gold nobles.

Area: Ipswich, Suffolk. Found: 1863. Buried: 980.

This hoard of 500 silver coins was buried before, and escaped, Danish pillaging raids in A.D. 991 and 1000. It lay buried for almost 900 years unprotected by any kind of container, yet about one hundred of the coins, mostly silver pennies, were in good condition. It is thought the acidic soil contributed largely to their preservation.

Area: Kinghorn, Fifeshire. Found: 1864. Buried: 1350.

More than 1,000 silver pennies and sterlings were in this hoard, another that had no container.

Area: Llanarmon, Denbighshire. Found: 1845. Buried: 1480.

A mystery hoard in every sense. It contained about 100 gold and some silver coinage. Shortly after it was discovered, the hoard vanished; the authorities never saw any of it again.

Area: Newbury, Berks. Found: 1756. Buried: 1307.

A big hoard (3,499 pieces) made up of English, Scottish and foreign coins, mostly silver pennies.

Area: Northampton, Northants. Found: 1873. Buried: 1290.

Records on this one are slightly blurred; they don't say whether it was loose or in any kind of container. It consisted of 194 Edward I silver coins, mostly pennies.

Area: York, Yorkshire. Found: 1704. Buried: 1070.

Classic hoard-finding situation: the hoard was in a wooden container seemingly discovered within the walls of a house when it was being pulled down. Inside the box was a collection of 250 silver pennies from the Norman period.

This very short list, with the barest details in it, is a guide to show what considerable treasure in coins and valuables must still be undiscovered throughout Britain. It does seem improbable that a container, or loose cluster of coins, could lie undetected for century after century.

How was it that somebody—a landowner, workman, serving maid, ostler, abbott, house dweller, forester, huntsman—on one of the tens of thousands of days it lay concealed, didn't stumble across it? A good question. A strong part of the answer must be the ingenuity and determination of the original owners or those who criminally obtained the valuables: they buried them to stay buried. Whatever the reasons, be assured that all the hoards so far discovered are only a tiny part of the number that still rest undisturbed, perhaps tantalisingly close to where you live, or have even been searching. If you have the wit, the enterprise, the systematic approach, the patience and the energy, your chances of one day finding a hoard are much higher than you could have previously believed.

I'll end this section on a high-value note with brief mention of one of Britain's most fabled, and puzzled-over, hoards: King John's treasure. A bounty of untold value, its presumed location has been narrowed down to a relatively small area: R. Nene's canal in Lincolnshire, not far from the Norfolk border. But, for more than seven and a half centuries the treasure has kept its secret; it is still presumed to be lying where it vanished in October 1216.

In those medieval days, the monarch had to devote much, perhaps too much, time and effort to keeping his treasure intact. There were three main repositories of the nation's wealth: the State Treasury, the Exchequer and the king's personal fortune. To make his valuables less vulnerable, the king dispersed them around various parts of the country, in widely scattered abbeys. He was also compelled to take a burdensome amount of it with him on his travels, which were many. So, accompanying king and court wherever they went was a caravan of cumbersome carts, creaking and lurching along at a couple of miles an hour.

Along the way, the king would sometimes break off from the baggage train to check on his hoards; the carts would continue to their destination along the most direct route.

Late in September 1216, King John banqueted, far too well as it proved, in King's Lynn. From there he chose to travel through Wisbech and Spalding to Swineshead; his baggage train was to follow, making its slow way across the Wash estuary, a journey only possible at low tide.

Perhaps its progress was much slower than calculated. The tide rushed back in, trapped the laden carts and they sank forever into the quicksands. The king had no chance to organise an immediate search for the baggage train. The feasting at King's Lynn afflicted him with an intense bout of dysentery. He died before the year was out at Newark in Nottinghamshire, aged forty-nine.

Nobody could estimate just how much treasure was sucked under the sands, but its bulk must have been considerable. Away from human gaze and knowledge went King John's superb jewel collection, his ornamented plate, a profusion of silver and gold flagons and goblets, bracelets, rings, charms, candelabra, his coronation robes and regalia, his great crown, the golden wand, the golden dove and the jewel-encrusted Sword of Tristam.

Since then scholars, historians, adventurers, treasure-hunters ancient and modern have pondered the mystery. But not a single authenticated item has yet been recovered. King John's Cup, the earliest surviving piece of English medieval secular plate, is now on display at the Guildhall, King's Lynn. At first it was thought to have been part of the missing treasure, but experts established that it was fashioned in the following century.

One major problem for the treasure-seekers is that they have not been able to chart the exact route of the ill-fated baggage train. Whatever the path, it now lies many miles inland. Technology is the increasingly efficient servant of archaeologists; its sensitive instruments may be moving them significantly closer to solving this elusive mystery.

In recent years, research teams from Cambridge University have made two protracted bids to locate the treasure. The first was launched in 1959 and the second in 1966. Searching was spasmodic rather than constant, and covered some three years altogether.

The searchers mapped out the likely area with aerial photographs and did a series of test borings, going down forty feet on occasions. Twenty feet down one bore hole, they came across fragments of chain mail, chips of gold and pieces of metal, presumed to be sword shards.

It would need the bankroll of a Howard Hughes to finance another search that would be anywhere near adequate; the capital needed would run into hundreds of thousands of pounds. If such a search was completely successful, and the entire hoard was located, the lucky finder would scarcely have an opportunity to caress a single item before the whole lot would be whisked away from him. The authorities would bundle it up, surround it with the most massive security guard ever mustered in British history and transport it direct to Queen Elizabeth II or whoever was occupying the throne at the time. It would be the property of King John's reigning direct descendant.

The most delicious thought about King John's treasure is this: does it in fact lie buried under the sands of the Wash at all? Or was the story of its loss an elaborate ruse to deceive the Ronald Biggses of the time and frustrate the king's heirs?

King John, art thou smirking there below?

CHAPTER 6

SEARCHING FOR HOARDS AND OTHER OBJECTS IN COMMONS, FIELDS, FARMLAND, FORESTS, PONDS AND ON FOOTPATHS

AT the start of the previous chapter I listed some of the many places where hoards can be found. Among them are woods, copses, farmlands, commons and footpaths. Of course, you visit these places to look for many other things besides hoards; the following pages will tell you what these items are, where they are likely to be and how to recover them.

However, before focusing on less glittering finds, I do want to delve into history to explain why so many hoards have been squirrelled away and why so few of their number have so far been recovered. To do this, I have chosen to quote in full an extract from a very old and most fascinating book.

This weathered tome contains two bound volumes of *The Universal Magazine of Knowledge and Pleasure for 1784*; it is the kind of book you should now be seeking as a reference source. The extract is sparely presented; it is simply a list of the robberies committed by a young, and unexceptional, thief and highwayman of the eighteenth century. It conveys the physical dangers and harsh justice of those times and, most importantly for us, presents a mystery: the unexplained whereabouts of money and valuables, i.e. potential hoards.

Here is the extract:

August 23, 1784

Account of the Robberies committed by Joseph Radley, not eighteen years of age, who was executed at Aylesbury on Thursday, August 5.

	£	*s.*	*d.*
1. From Councillor Dallas, in Kensington Gardens	4	14	6
2. Earl of Buckinghamshire, ditto	36	15	6
3. A clergyman and his lady, ditto	7	7	0
4. An old gentleman, ditto	1	8	0
5. Another person, ditto, a watch, and 2*s*., returned			
6. Barnet-rd, two ladies in a post-chaise	26	5	0

		£	*s.*	*d.*
7.	Ditto, two ladies and one gentleman in a post-chaise coming to London, two twenty pound notes, 17 guineas and a half, 19 shillings, two Spanish dollars, three pocket pieces, a lady's watch, a silver one, and some trinkets, amounting, in the whole, to about	75	0	0
8.	Between Twyford and Reading, two gentlemen and three ladies in a coach	5	5	0
9.	Near Bagshot-heath, a gentleman and a lady, in a post-chaise, a watch &c.	12	12	0
10.	Mr ——, of Argyll-st, and his lady, in a post-chaise	17	5	0
11.	A Quaker, in a post-chaise, near Henley, of a watch &c.	3	3	0
12.	Near Bagshot-heath, a gentleman in a post-chaise	3	3	0
13.	Two foreigners, in a post-chaise, near the Bath road	16	6	0
14.	Forged upon a certain Lady, in keeping by Lord ——, for	17	0	0
15.	Feather-bed lane, near Stoken-church, three ladies in a post-chaise, an enamelled ring, one gold ditto and 10*s*., in the whole	1	5	0
16.	A man, near Epsom, of		2	6
17.	Near Kennington-common, a gentleman and two youths, in a post-chaise, in gold and silver about	3	0	0
18.	Barnet road, from a gentleman and two ladies, in a coach, a gold watch &c.	4	14	6
19.	Epping-forest, a gentleman and lady in a coach	12	12	0
20.	Ditto, the same morning, an old clergyman	2	10	0
21.	Remnan-common, near Henley, four officers in a post-chaise	19	19	0
22.	Park-lane, Hyde Park, a gentleman in a coach	7	7	0
23.	An old lady, in a coach, near Dunstable	26	5	0
24.	Bath road, near Burnham, for which he was tried and condemned, for robbing two ladies of a watch, with a gold seal, and a small picture, a ring, and one guinea	4	0	0
		£308	8	6

After his being apprehended for this last robbery, he made his escape and was at large five weeks, during which time he committed six trifling robberies, and also several others, he said, not worth mentioning.

The magazine did not say what became of the money and valuables young Radley stole. Were the authorities able to recover most, or any, of the stolen loot? Was Radley able to retrieve and dispose of whatever was left during his final five weeks of freedom? Did he put it for relative safe-keeping in one central hoard, inside a house or in the open, or divide it into a number of secret caches? If he split the stuff up, did he perhaps forget the location or locations of some? A simple enough thing to do.

The fate of this lad, strung up to die before he was eighteen years old, seems pitifully cruel and grotesque in the light of twentieth-century laws and attitudes. But in the eighteenth and nineteenth centuries there were hundreds like him. A small number of fictional counterparts were immortalised as Fagin's gang of pickpockets in *Oliver Twist*. Social and working conditions thrust youngsters and adults, most of them illiterate, into summing up their lot as starvation or crime. So the cities' mean streets and alleys seethed with Bill Sikes-type burglars, pickpockets, thugs and cutpurses; and behind the dark, rustling foliage alongside so many roads waited the masked and caped figure on horseback with a primed flintlock pistol in his hand.

Privation bred crime, as it always had, and crime meant more and more hoards secreted away by those anxious to keep their possessions and those who had done some dispossessing.

It's hard to perceive just how much basic insecurity has ebbed from the lives of ordinary people in the last three or four generations. Nowadays we can sit snug in our homes with the services of a large and immensely sophisticated police force a telephone call away. To protect our valuables we can have a host of devices: triple door bolts, burglar alarms, tumbler locks, home safes, electronic surveillance systems, etc. If calamity strikes, we have insurance against fire and theft to recoup our losses. The person who thinks all this is still not enough keeps his valuables in a safe deposit box at his bank.

For most people, none of these layers of defence existed before the twentieth century.

Before then, people had to devise whatever means they could to guard their possessions. The first crude coins were made and circulated in Britain more than 2,000 years ago; gold was valued and feverishly sought long before then. But the first banks—and they were restricted to a privileged few—were not launched until late in the eighteenth century; banks with services for the general public about one hundred years later.

So yesterday's Britons were afflicted with a fear, which is less harsh for us today, thanks to modern safeguards: the fear of intruders wrenching away or silently removing every single valuable they owned. The shrewd concealed their belongings with cool deliberation; the unwary in trembling haste. Imagine the panic among those ancient Britons who heard of the impending arrival of Viking, Saxon and Norman invaders or who heard the drums and saw the sunlight flashing on the breastplates of approaching Roman legions. Their few bits of plate and trinkets were scooped up, bundled together and stuffed into the nearest hiding-place—down a well, in a knot-hole of a tree, under straw in the barn, anywhere.

The Danish battle-axe, the short Roman sword have been a random threat. A more potent menace has always been present: the tongues of neighbours and fellow villagers. Until the Industrial Revolution began to tug people from farms and tiny settlements into cities that gave them anonymity, almost everybody lived isolated or in a small community. The threads of their daily lives were tightly woven with their neighbours'; secrets were harder to keep.

The entire village knew who was the local miser and pondered how much money he had and where he kept it. Villagers also mused over what Farmer Tressider did to guard the coins that jingled in his pocket as he strolled cheerfully home from selling his cows and pigs at the market. Both miser and farmer were alert to such curiosity and made sure it wasn't too easily satisfied.

Many, of course, had brief contact with their money. For instance, in the last century, a gentleman farmer from Ipswich was held up by a highwaymen not far from that city. The masked fellow relieved him of four hundred golden guineas and one Portuguese piece. (A modern stick-up man or mugger seeking a similarly ripe prospect today may find in his wallet £3·50 in cash and a fistful of credit cards.) Yet, these enormous and massively heavy loads of coins in bygone times weighed down the pockets of many merchants and travellers. They were able to achieve a better balance and sartorial line on foot or horseback when the first money belts were introduced in William IV's reign, first as a dandy's fancy, then as a practical addition to the sensible man's travelling outfit.

For many the hole beneath the stone-flagged floor or behind the loose brick in the fireplace was too obvious a hiding-place. So they took the bulging wash-leather purse, the casket, the urn, the biscuit tin and buried it elsewhere.

A favourite hiding-place was in a clearing in a near-by wood. Not all that imaginative but somehow safer than leaving the groats, sovereigns, the silverware inside the dwelling-place. So, around Britain through hundreds and hundreds of years the same furtive sequence has been played out over and over: the silent figure moving through the darkness to the chosen spot; the stealthy thump and rasp of shovel against earth; the laboured breathing; the sigh of satisfaction and relief; the careful journey back to the cottage or hovel or farmhouse.

And in some such fine and private place almost everyone of those hoards remains to this day. There must be thousands upon thousands of them. The man who put the stuff there could have buried it so hurriedly he forgot exactly where it lay. Death came early and swiftly in a multitude of ways: from infection, from plague, from wounds of battle and for some—like Joseph Radley—from the rope. Surely more than a few of Radley's fellow-thieves of different eras must have spent six months huddled in the stinking holds of convict ships bound for Botany Bay, thinking wistfully of a small box resting beneath the grass of some quiet Hampshire glade or rocky Norfolk moorland. So many hoards remain intact because only one person ever knew where it was—and is.

Another vital reason for non-discovery is that few people until recent years have deliberately and systematically set about locating the hoards. It would be most surprising if more than a thousand, a small fraction of those existing, have been discovered in the last three hundred years. Your chances of finding one are better than ever.

But only if you go about it in a logical and enterprising manner. This means putting considerable effort into the task, both in research and searching with a patience to make Job seem to have been a restless weakling. Extensive preparation is more vital for seeking buried hoards than any other kind of treasure-hunting; you just can't shirk it. Tracking down that hoard can take months of piecing together many clues that can prove maddeningly elusive. You'll find them in what should now be your accustomed haunts: books in the local library, the Public Records Office, parish register and newspaper office. From old books and almanacs you'll learn of eccentrics, known or suspected misers, yesterday's men of wealth and station, of robbers and robberies. Records will tell you of men whose wills and bequests, the estates they left do not correspond with earlier stories of their means and property. Parish registers of the eighteenth and early nineteenth centuries could be startlingly candid; if a man was thought to be a miser and misanthrope, the local clergyman often said so bluntly when recording his death. If you can get access to the cutting library of your local newspaper office, you'll find that its sections on 'Hoards', 'Treasure', 'Old Buildings', etc. may contain promising nuggets of information. Their old newspaper files may provide useful details about the times before their cuttings were first collected.

To this research you can add talking with elderly people of any district that interests you. Many of their stories will be handed-down gossip; others with a hard kernel of fact will have been fancifully embroidered down the years. So useful leads must be cross-checked.

If you live in a village or small town, it will have been less ravaged by the wreckers' ball; a good percentage of old buildings should still be standing. Note the fine old houses, the sturdy inns with large cobbled courtyards where horse-drawn carriages used to stop. Learn who has lived in them and what happened to those people. They all had their secrets; those that have leaked out, prompting rumour and conjecture, could set you on the path towards a hoard.

The innkeeper could have been in league with a highwayman, or received a steady income of sovereigns by letting the local squire use a cosy back room as a trysting place with his mistress. The outlaw's booty and those gold pieces may still be near at hand.

The most practical course is to research places near where you live. You can keep going back to your sources of reference. And, once you have selected a likely site, you can revisit it time and again. This may be necessary; in fact, you would be amazingly fortunate to locate a hoard on your first day of searching; the football pools offer shorter odds for success.

You won't have to seek far and wide. Every county of Britain abounds in forests, copses and dells that could conceal a hoard. Many of these are on the fringes of towns and villages or in corners of farmlands; bygone residents could reach them easily. They had to because many hoards were, in fact, private banks to be replenished and taken from repeatedly. This need for easy access also explains why most of the hoards so far discovered in

Britain and elsewhere, have been less than three feet under the soil.

How far you'll have to probe down depends very much on the nature of the terrain. If the ground slopes, a hoard could be very close to the surface; erosion after rainfalls will have carried earth and leaf mould down the slope. Therefore, on the lower reaches the accumulation will mean your target is deep down—perhaps ten or twelve feet.

In pirate movies, the area containing the so-familiar metal-studded chest was often clearly marked; that allowed the film-makers to keep plot and action moving. In real life that didn't happen. The marker obvious to the burier was also obvious to his curious neighbours. So don't waste time seeking a cairn of stones, a tree with a cross or symbols gouged in its trunk.

If the hoard-owner was naïve or foolish enough—or his burial-place was remote enough—to have put up a marker or bury his box at the foot of the only obvious tree, the signs will almost certainly have vanished. A storm might have uprooted the tree, lightning could have obliterated it or foresters felled it.

Clearings are the best places to search. People did seek them out as burial spots; they offer ample room to move around; and they are clearly designated areas you can work from timberline to timberline. Experienced treasure-hunters very confident of their sense of terrain could try to work them eyes-only, but this is not really recommended. You should set down your constant companions, the line and pins.

If the clearing is very large, it's best to mark out the usual fifteen-foot squares. If it's quite small, say approximately forty feet long, lay down the lines along its entire length, work this lane in one complete sweep, move the pins across, then come back again. You've gone into the woods purposefully looking for something, so don't be slapdash. Make sure the detector head doesn't miss one single square inch; search intently in clumps of grass, inside and underneath bushes. If you expect to come across brambles and nettles, have a pair of thick gloves to protect your hands. Searching woods is the perfect way to learn patience, so it may be less daunting to begin with a small wood that your researches have labelled 'promising'. Choose one you can cover in a month of regular searching. And, never forget you may have to spend weeks or months working a single copse or wood. If you can endure this, you have the makings of a good treasure-hunter, and not just a week-end dilettante (Oxford Dictionary definition: trifling, not thorough, amateur) with a detector.

I advised earlier against focusing too hard on a gnarled and misshapen tree or any irregular features of the copse as the likely place for a hoard. If you are systematic, you will check them as a matter of course. Snoop the detector head like a hunting dog's nose around all the roots. Then move the search head carefully upwards along the trunk and into the branches, as far as you can reach. If you have a light folding stool strong enough to take your weight, bring this with you to stand on and get your detector head as far into the tree as possible.

The miser, footpad or highwayman may have climbed the tree to place the casket, purse or whatever into a fork or knot-hole well above the ground. A man either sitting or standing on his horse's back could have reached up a long way.

In woods close to habitation, there is almost as great a variety of items as you would find on a public common: old coins, pendants, brooches, small purses, rings, watches, buckles. In woods, as in open country, you can find any number of offbeat items. For instance, religious medallions, which are becoming more and more popular among collectors, are found on all manner of sites throughout the entire country. Wherever pilgrims have trod, they have dropped medallions; they also tossed a great many of them into streams and rivers, precursors of those people who can't resist hurling coins from any kind of bridge.

Other things that were obviously lost—they were found at too shallow a depth to have been deliberately buried—need more explaining away. How did a large and very ornate silver picture-frame that one treasure-hunter found, happen to get lost in the woods?

A baffling question-mark also hovers over such things as fire-backs, masses of insurance plaques, an elaborate brass plaque of Eastern origin and tradesmen's measuring weights that have all turned up in rural places. Some of the weights, made of lead, go back to medieval times. Later ones, from the nineteenth century onwards, are of brass, hallmarked, dated and often bear the merchant's name. These are now much sought-after and bring high prices. As you search more and more, your sense of surprise will shrink: the most bewilderingly unlikely things will keep on emerging from the ground in the most unlikely places.

OPEN COUNTRYSIDE AND FARMS

Farms can be excellent places for searching. The fields themselves have their attraction; many hoards and ancient artifacts have been discovered when a shovel or ploughblade has clanged against them. Enormous areas of Britain have been cleared, first for cultivation, then for urban development in the past century and a half. Yesterday's woods and copses, that stayed largely undisturbed for thousands of years, are the fields of today. Yet the immense size of even small fields, compared with other sites, make them impractical to search unless you have very powerful clues to something specific buried under them.

Marking out fifteen-foot squares in a field several acres in size may induce the sort of hopelessness you would suffer trying to wear away the Rock of Gibraltar with a nail-file. You also run the risk of disturbing ancient sites.

You could try a field where ploughing has brought to the surface tell-tale signs of white clay and perhaps red and black chips of broken pottery; all these would be nicely visible after a good rain. The pieces could denote that it's an old Roman field that may yield new stuff. You would be best advised to tell the local archaeological society about it and let the members do the work. Open farmlands are possibly the least productive sites of all, and they could be of historical importance.

Farm buildings and their environs are a different matter. They are most promising for both lost and buried items. You have the choice of the farmhouse, barns, outhouses, approach roads and the inevitable network of footpaths or farmworkers' trails. Most of the major footpaths of Britain were tramped upon more than 1,000 years ago; the same applies to old farms.

Many generations of the same family have been working their farms for centuries. Long ago they established their rubbish-tips. Into these they have accidentally tossed household utensils or deliberately hurled their worn-out implements, a lot of which were made of metal. You may come across surprises, like a six-foot and 100-year-old plough in perfect condition that one treasure-hunter retrieved from a farm rubbish-tip.

Farm pickings can be good because their seven-days-a-week work schedule has never allowed farmers spare time for such secondary pursuits as treasure-hunting. So a detector can be a fine gift for a farmer's son. Even if his own spare time is limited, he can allocate some of his former fishing and shooting time to searching and quickly build a good collection.

If you have a farmer friend, ask permission to search the farmhouse from top to bottom and the other buildings as well. He may be startled at the amount of stuff you find.

COMMONS

Next to working your own living-room or garden, commons are the simplest areas to reach for detecting. More through luck than foresight, commons are generously sprinkled throughout many of Britain's large cities; every town and village has one or several. They are the historic outdoor meeting-places; virtually every act that men carry out on terra firma has taken place on commons, from conception to death in a multitude of forms. Throughout centuries, people have frequented and gathered on commons, for moonlit romances or robberies, for promenading, to watch bear-fights, cock-fights, bare knuckle bouts, for the colourful or sleazy fairs that still liven them with their gaudy sideshows, whirling Ferris wheels and dodgem cars. You'll learn something of their histories in your library.

If you stand on a common anywhere and sweep your gaze around its grassy expanse, it is almost impossible to comprehend the amount of lost money and valuables that lie beneath them. Diligent and regular working of such areas could bring you premature curvature of the spine from stooping to gather your finds.

Documents will tell you where yesterday's generations gathered on the common; ten minutes looking around the area and the people within it, will guide you to the fruitful locations.

Don't say to yourself: 'This is a pretty big place. There must be so much stuff under the ground; I'll just wander around here and there.' You have to work there as slowly, carefully and systematically as you would in any other place. One veteran treasure-hunter once wagered that he could find more on Hampstead Heath than six relative novices who were with him. They spread out in a line and covered fifteen or twenty yards of terrain. He followed behind them, covering exactly the same area. At the end of an hour, his haul was five times their collective finds.

He believes their weakness, apart from allowing their detector heads to wander, was possibly the most widespread among detector users: keeping the detector head too far from the ground. (Since then Hampstead Heath has been closed to treasure-hunters.)

Unless you keep the detector head less than half an inch above the ground, you have wasted your money buying it (Fig. 6—example at right). If a mole stuck just his nose from a hole in the ground you are working, the underside of your detector should brush against his nostrils. Anyone taking up golf, tennis or

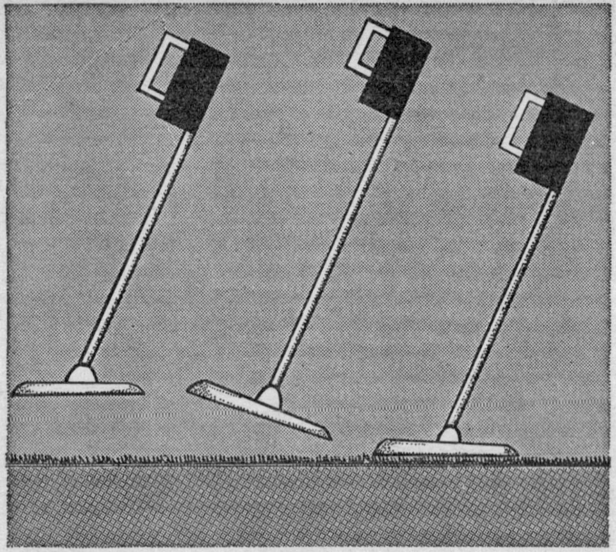

Fig. 6

cricket has to spend time practising his strokes. The same applies to detecting. At first, be conscious of keeping the detector head low, low, low and of sweeping it back and forth in slow, straight lines. A surprisingly common fault among fledgling treasure-hunters is to wave the detector like a pendulum. That may sound improbable but it happens all too often.

The first thing to remember is that your grip must be firm and relaxed. Experiment with your grip until you achieve one that allows you to work the detector with an outstretched arm—which is by far the best way—while keeping the head low. A shiny plastic handle can become slippery with perspiration. It's best to bind insulating tape around the handle. This may be somewhat sticky at first, but will settle to give you a secure, non-slip grip. Rubber handles need no treatment.

You'll find the outstretched arm is the least fatiguing position and allows you to sway the head from side to side in a light and easy movement. The other main point on technique is to keep the detector head always parallel to the ground. If you let it tilt at all, you lose some or all of its penetration. (Fig. 6.) Where the ground slopes—on tussocks or the edges of deeply rutted paths—adjust the head to keep it parallel to the lie of the land. Like keeping your eye on the ball when playing golf or tennis or squash, for example, this is a basic rule some people are apt to forget.

So you're working a common, and the detector emits a note that tells you something metallic is under the surface. If the sound suggests a small object about six or seven inches down (experience will train your ear), you can get at it by splitting the grass with a screwdriver tip or knife blade in the shape of a cross (Fig. 7—for example at left). Insert the screwdriver shaft down into the intersection of the cross to probe for and flick out the object. Don't be exasperated if it turns out to be silver paper or a can ring-pull; the world is full of thirsty people who like to munch chocolate or smoke cigarettes.

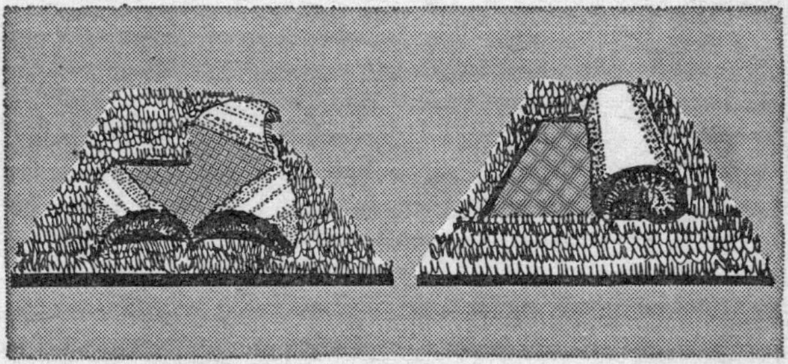

Fig. 7

For a deeper probe, cut out three sides of a square (Fig. 7—example at right). Usually a square about three or four inches across is large enough. Slide the knife blade or trowel underneath and hinge back this flap of turf and soil. Be careful not to flip the square divot back so hard that it comes free from the ground. Then check the flap with your detector to find out if the object is there or still resting in the ground.

If the detector reading is strong, indicating something large like a horseshoe, cut out and lift back a square up to nine inches across. It's better to cut a small square, then enlarge it if necessary, than to gouge large ones at first. While the trowel blade is curved, it will cut a pretty straight line if you do it carefully. Once you've removed the object, push the divot back into place and stamp it down hard with the sole of your shoe.

Some instruction books have recommended removing the plug of earth entirely from the ground. This is absolutely wrong. The grass on the plug will then wither and the green surface of the common will be speckled with faded brown patches that indicate an incompetent, and foolish, searcher has been at work.

If you dig the wrong way, you will blight more than the patches of grass; you'll be undermining the pastime of treasure-hunting. After a series of complaints from not-disinterested parties, the Greater London Council in 1972 began declaring public commons out of bounds to detector users. This ban was extended until virtually all GLC parklands were forbidden territory. Similar prohibitions came into force in a few other parts of the country, such as Ipswich in Suffolk.

Unless every detector user takes the greatest care with every hole he digs, all commons around the country will be systematically closed to them. Treasure-hunters must realise they are very vulnerable. Nobody can, or should, take offence at people who dig on commons sensibly enough to make sure they leave no traces of their visits. After all, dogs that uproot grass and bushes and scrape holes in the ground mar the landscape much more. Yet there are no by-laws compelling dog owners to repair the damage that Fido or Glenwallis Fitzwilliam IV, the award-winning thoroughbred retriever, inflicts on public land. Most individuals and authorities in Britain are reasonable and tolerant towards common-sense behaviour. So please don't give anyone an excuse to crack down on treasure-hunters. As a matter of course, make sure that all the divots you've prised up look, when replaced, as if the turf had not been disturbed at all. This takes no more time or effort than slapdash and negligent work.

If you're still a beginner, the footpaths that criss-cross commons, fields and farmlands are ideal for gaining experience. In fact, after you've worked over your house and garden, they should be the next sites to search. They have all these good points: they have exact and clear boundaries, they are well and widely used, they are easy to find and very close to where you live because they are found in abundance in every county of Britain.

Now is the time to bring your ordnance survey maps into action. The 1874 series will tell you where the older paths are; newer maps show new paths and those that have been at all altered during the last century. Every footpath in Britain has been mapped, so you have access to an invaluable guide wherever you live. And you have a right under law to tread the majority of them. For instance, a public path runs through the land on which Chequers, the Prime Minister's country home is sited. We don't recommend you try to work it; anyone wearing headphones and carrying an electronic device in that area would quickly find himself surrounded by Secret Servicemen and moved on—or worse.

You should work the entire width of the path and a margin up to eighteen inches on either side (Fig. 8). These margin areas are important because many people have walked close to their sides. In rainy weather when the paths are muddy, they often choose to walk alongside rather than on the path. You can coinshoot very

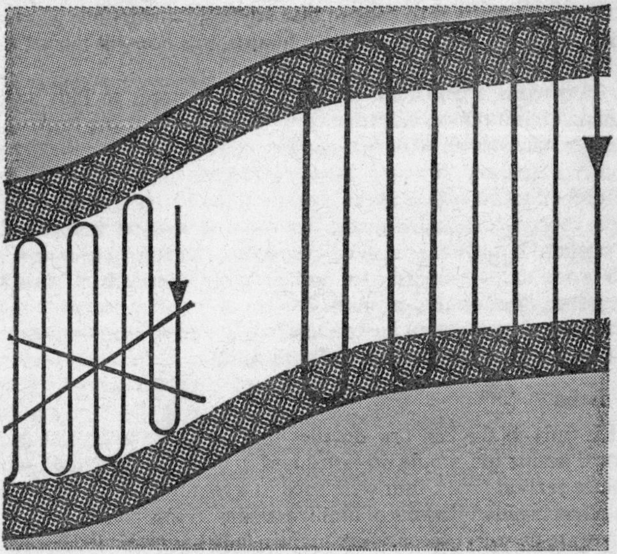

Fig. 8

profitably on modern footpaths, but older ones are better for the wider range and higher value of things you are likely to find. As you work more and more of them, you will realise that some of your best discoveries will be under footpaths worn completely flat by thousands of feet down decades or centuries.

Many footpaths either have been worn down to hard pack gravel or have a layer of gravel very near to the underclay beneath them, so a trowel is your best implement for retrieval. As with trowel work on grassy terrain, a three- to five-inches-wide plug should be adequate for prising out most objects. This time, however, you can remove the entire plug. When you've run the detector over both the plug and the ground you gouged it from, replace the plug, and make the ground level again by stamping it hard several times with your foot. You can cover virtually all traces of your work.

Footpaths are not the most likely places for hoards to have been buried, but several have been discovered under them, both loose and in containers. One was found in an eathenware pot under a public path running through private land in Hertfordshire. The path cut through scrub and woodland with no marking or unusual features to suggest that anything worthwhile was concealed there. Like most footpath hoards, this one was buried very shallowly—it was only six inches down.

Footpaths are in fact unusually clever places to hide something. Operating on a double cunning, something like bluffing at poker, the owners reasoned most people would not consider an open place trod daily by dozens or hundreds of pairs of feet a safe hiding-place. Admittedly, retrieval for the burier was more difficult than in a secluded glade, but the ruse worked well enough. Footpath hoards are indeed a bonus because you go to footpaths to work them primarily for lost articles. However, if your researches disclose that a miser, recluse or wealthy eccentric lived in the area, or if olden-day robbers frequented it, keep the thought of such a hoard in the back of your mind.

PONDS

It is only in the last few decades that ponds have ceased to be focal points for people on farms and in villages and small towns. Since earliest times, they were used as a source of drinking water; women brought baskets of family laundry to wash on their banks; they are trysting places where many a lady has unpinned her hair; thieves and untrusting folk have dropped their hoards into them. In this century, criminals, for some reason, have thought them more enticing hiding-places than did their dishonourable forerunners. Coins, jewellery, metallic objects don't have to be waterproofed; leather pouches can let in water without harming their contents unduly. A favourite trick was to lash twine around the container, toss it into the pond, then tie the loose end to a bush, tree-trunk or roots, and conceal the twine under lily pads or fronds. Thus the cache was lost from sight. The twine could last for years.

Knowing the types of people who've had constant access to the pond, you can search and expect to find articles both in the water and on the fringe of the pond. Wearing wellingtons or waders, rake the bed carefully and use your detector on the banks. You can have blank days and be infuriated by the amount of rusted metal rubbish your rake will snag and draw up. So, if there are several small ponds in a small area, try to work a few in a single day; they can be done quickly. They can also offer up a host of excellent and varied finds, being good sources of old and more modern coins.

SEEKING PERMISSION

Whether you plan to work commons, fields, woods, ponds or footpaths, always make sure you have either a legal right or specially sought permission to go to the site. All land belongs to somebody. Maps will give some guidance as to ownership and right of access. It's simple enough to find out what is National Trust land, Crown land, or where archaeological sites are situated.

If you approach a farmer or any other kind of landowner for permission, always project a sense of responsibility. Be perfectly frank and hold nothing back. If you are seeking a bucket of gold coins that research had indicated should be on that land, or merely want to search generally, say so explicitly. And emphasise that you will leave the place exactly as you found it—no gaping holes, no mounds of dirt, no mess. It's quite common for those unacquainted with treasure-hunting to suppose you are going to dig a hole in which a double-decker bus could be buried, even if you propose to look for something no larger than a biscuit tin. Approach the owner in the proper way and you are almost certain to obtain permission, although he may consider you somewhat eccentric.

If your attitude is right, it would also be most unusual if the owner insisted that you hand over all you find, or even split it with him.

AMMUNITION

I said in the opening chapter that treasure-hunting has helped to spread the mania for collecting an immense variety of things. This applies especially to the collection of many types of ammunition, an activity that gathered momentum in the early 1970s. A lot of people began to go out and look specifically for ammunition, notably in London's Home Counties.

This happened because treasure-hunters who had gone out coinshooting or generally searching open areas found they were coming across lots of Georgian musket balls and a startling variety of Victorian ammunition. Often these would be found in the same areas, not surprisingly when you consider the natural thread of continuity in the military forces; Victorian camps carried on from Georgian camps and so on. These areas also yielded plenty of ammunition used between the Boer War and the outbreak of World War I.

There are many former army installations in forests that were used for long periods or bivouacs where troops were quartered during exercises that may have lasted for a few days or several weeks. In many of these places, ammunition was buried and not retrieved.

For ammunition hunters, there are plenty of references. You can find good clues in regimental histories and also in newspaper files. Local papers have carried stories of exercises and complaints aboutt he noise from firing ranges, the conduct in streets of troops on leave, about soldiers brawling among themselves or fighting civilians in inns and eating-houses. Those ever-useful ordnance survey maps of the 1870s are perhaps the best quick references. In the last century, military matters were far less hush-hush than they are today. If there was an army camp in the area, it was marked on the map. If you study both old and modern maps, it's quite likely that the highest point marked in the district had an army installation on it at some time.

The abundance of Victorian ammunition, after all these years, is not surprising; a great deal of experimental work on munitions was carried out during the nineteenth century. After the musket ball gave way to the cartridge, every year seemed to bring a new variation in bullets. As happens in all kinds of research, grotesque devices were fashioned, tested, discarded and, then, it was back to the drawing-board for the research teams. One extraordinary rifle-cartridge was three inches long and one and a quarter inches across; it looked more like a miniature cannon-shell. The most sensible and humane thing to have done with these would have been to have put a few in a gift package and sent it by courier to the enemy with a note reading: 'This is the sort of thing we have at our disposal now. Why not give up?' Instead it was tested on rifle ranges, where the recoil fractured shoulder-blades and snapped the spines of hapless riflemen.

Among the most popular items sought are artillery shells. People search at opposite ends of areas near old ranges for the metal casings and nose cones. The dangers in hunting for this stuff are self-evident but the lure is strong: the casings, polished to a high gloss, look very handsome perched on mantelpieces or fashioned into lampstands, pedestal ash-trays and other things.

To all treasure-hunters, even those who aren't seeking such items, we give this advice: know your neighbourhood or search area as well as possible so that you're on the alert for possible ammunition finds, above or under the ground. You can stumble across the most unexpected things. One treasure-hunter searching an Essex wood discovered a cache of Bren guns so rusted and corroded that they were almost pulped metal. When he reported the find, his local police were amazed. Even in the less frantic atmosphere of peacetime manoeuvres, the movement of hundreds of personnel and tons of equipment can never be perfectly co-ordinated; hence the myriad of lost articles, from pannikins to tool kits, and the forgotten buried caches of ammunition. (During World War I, a slipshod filing clerk was said to have somehow mislaid an eighty-ton steam locomotive.)

When you do locate ammunition, common sense must prevail; regard it as you would any potentially lethal object. If you suspect it could still be alive, *don't touch it*; certainly don't lay a finger on such things as hand-grenades and mines. Mark the spot very clearly, jot down in your notebook exact and unmistakable reference points, then fetch the police or army authorities. Foolishness and tragedy could very well lead to a tightening of laws on areas of access, or possession of any kind of ammunition, even palpably dead stuff, and another segment of treasure-hunting would be blocked off.

Firearms and munitions excite and fascinate youngsters, who have succumbed to the glamour of war presented in their adventure books and countless movies. They are naturally attracted to collecting such things as shells and musket balls. Parents buying detectors for their teenage children should spell out to them the dangers of ammunition hunting. For many reasons, it's important that only responsible people take up treasure-hunting; the pastime can well do without the other kind.

CHAPTER 7

RIVERS AND TIDAL ESTUARIES: THE TREASURE-HUNTER'S MOST PRODUCTIVE SITES

FOR treasure-hunters, a tidal river is like a conjuror's wineglass: no matter how much you take from it, it is never emptied. Two zealous treasure-hunters have been working the same 200-yard urban stretch of the Thames between them since 1940. On each visit they swoop on more finds and they know their patch will still be yielding in the year 2000, or even 3000 if the planet sustains any life at all then.

And there are thousands of productive sites like this throughout Britain. Every new tide floats in or disturbs its cargo of debris and valuables. That's why any tidal river will always be your most fruitful site. Nature, in the relentless movement of water, is abetting you. So are history and man's splendidly clumsy habit of losing things. Losses were often deliberate; the Romans tossed coins into every river before crossing it; this appeased their gods.

Early man—looking like King Kong's younger brother—used rivers aeons before he learned to fashion his crudest implement, drape rough skins over his nakedness or blundered across the means of making fire. From the beginning, water sustained his life, from the slobbering mouthfuls he gulped for survival to his stalking of distracted animals drinking wherever nature allowed water to gather.

Throughout all those millenia, man has been in and near water, as he has waged battles, romped and frolicked, made love, carried on trade and transported goods and other fellow-beings. He has raised structures alongside and over bodies of water and, in rage and despair, killed others and himself in them. While carrying on all these activities, he has lost, mislaid and hidden things from rough flint axe-heads to this year's shiny new 50p piece.

Carelessness and gravity have combined to pull down to river beds items dropped inevitably from every bridge, every building, towpath, grassy bank, barge, tanker, raft and excursion boat.

Nature has helped in her violent way, collapsing bluffs and whipping up storms to tear apart buildings and bridges; to suck all manner of vessels, from coracles to oil-tankers, under frothing

Professional treasure hunter, John Webb, rakes the Thames near Blackfriars – a most productive site.

Another professional, Tony Hammond, searches a favourite Sussex beach with his two daughters. (*Photograph by courtesy of The News Centre, Portsmouth*).

Antique Bottles. At left of upper shelf are Codd's bottles with stoppers; below are ginger beer and hot water bottles.

Antique bottles come in hundreds of shapes and sizes; the rarest fetch £60 and upwards.

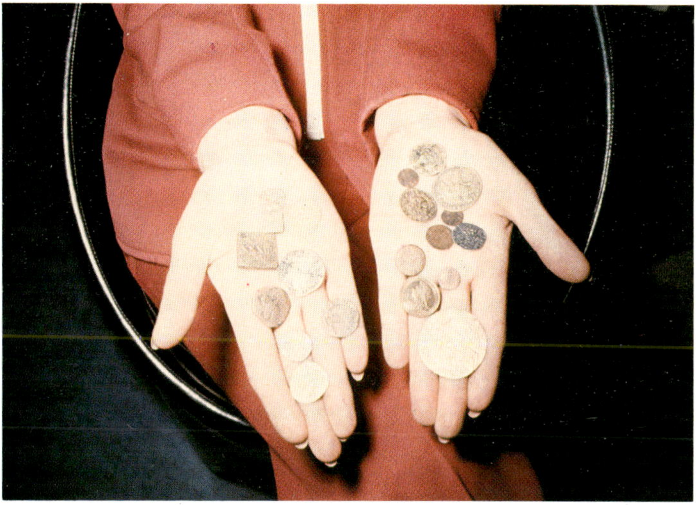

Two handsful of coins: Roman, medieval and modern. This lot are worth more than £1000.

The author admires a Unite, a 24-carat coin dated 1615. In good condition it would be worth £300 or more.

Old-style pot lids are in big demand.

An experienced prospector found all these coins and bottles in one day's searching. The red object is an underwater metal detector.

waves, and to send flood-waters to wipe out old landscapes and create new ones.

As history has unfolded, a veritable cascade of articles has plopped daily into rivers. At this moment, embedded in mud, sand or silt or being slowly stirred along by currents are millions upon millions of items: coins, pipes, pottery, bottles, purses, brooches, candlesticks, pot lids, and so on and on and on. Only a tiny—truly a microscopic—fraction of this immense cornucopia of articles has ever been recovered from British rivers; we still have only the haziest beginnings of a notion of just what is down there. The rest is waiting for the intelligent seeker to find. We hope you fit the bill.

You can take pot luck and go completely unprepared to search a tidal river in any part of the British Isles. Just toss aside your newspaper or girlfriend, put on your jacket and head for the nearest river's edge. Chances are you'll be rewarded with some kind of bounty from careful searching with only your eyes. Good luck if you are. But, in terms of productive all-round searching, this is like trying to learn architecture merely by staring at buildings, or expecting to qualify for a law degree by watching Perry Mason on television.

For best results, you must obviously begin with research. There are thousands of miles of riversides in Britain. Where are the most rational places to look for valuables in them? Start at your local library. Seek out charts, maps, photographs, documents, books on local history, geography, folklore and customs. Your library isn't likely to stock books on all of Britian's 300 tidal rivers, but should have a good selection of those in your district or county. If you want to know about rivers in Wales, Cornwall or the Orkney Islands, ask your librarian how to go about getting books on them.

And check whatever council records are available. From those, you can learn which different sections of your river or rivers have been or are most thickly populated, when industry began and what it comprised; you can learn about the river traffic, and about the warehouses, wharves, steps, towpaths, boathouses, seamen's dosshouses, etc. that once stood in the area. You'll want to know where people congregated for work, recreation and mischief, both amatory and criminal.

Where were the old fording places? Did the steel bridge that now carries four lanes of bumper-to-bumper traffic in peak hours replace a narrow-chested wooden foot-bridge built there in 1755? Has the river been dredged to allow deep-draughted freighters to steam along its reaches, and were the former banks hacked back to make space for quays and mooring-places? (Undredged and untouched sites are obviously full of promise.) What were the favourite dark stretches of riverside where the press gangs used to leap upon and stun with belaying pins clerks and homeward-bound shopkeepers who woke up in clipper ships bound for the East Indies? The miserable clerk's stickpin that popped from his cravat in the struggle could still be in the Thames mud. These are among the multitude of clues to the secrets the rivers conceal along their banks and beneath their gliding waters.

The most important piece of paper, or book, to study is the tide table you can buy from your local river authority. Just as the moon controls the tides, so the tides control your treasure-hunting sites. Note the daily high- and low-tide times, the periods of the year when spring tides are flowing.

You must know these things because you should time your visit to begin searching a tidal river site when the tide is ebbing.

Every stretch of river is unique, but two hours after high tide is a good average time to begin searching any river in Britain. You can follow the tide out as it recedes, then work back towards the high-water mark as it comes back. Allow yourself at the most four hours, which is as long as you can safely work any river in one session. If you have to walk out some way, as in estuaries, plan your exit route beforehand so that the incoming tide won't trap you. Seek advice from local residents wherever you can. Your quest may be engrossing—we hope it is—but pause from time to time to make sure the returning tide has not crept up on you. The sea will keep its rendezvous with the high-tide line no matter who or what is in its path. Respect its merciless power.

If you are the aesthetic type who wants his treasure-hunting to be a charming and poetic experience, keep well away from the most productive river sites, which are tidal and heavily industrial. The uglier, nastier and more forbidding the river area, the stronger the industrial blight that offends your eye, the more you will extract from that river. It can also be chilling and very mucky work. You can work sun-dappled streams in the Cotswolds, or sparkling burns in the Highlands, but you won't find one-twentieth of the valuables.

You want to search your chosen river seriously, so you must put some more preliminary effort into the task. Make a reconnaissance sortie. Take along a sketch-pad and pencil. From whatever vantage-point you choose (a high one is best), mark in the features you can see or have read about: wharf, old bond-house, windmill, factory, nineteenth-century picnic ground, disused steps.

If there's a bridge, stand in the centre and study the water carefully on the downstream side. Note the swirl of water as it tugs at the bridge piles, or washes around a prominent point. Study the river bends. The inner curves are the protected spots; the current is gentler here and small objects rest on the bottom because they can resist the pull of the water. Articles settle when their weight overcomes the current. This is why you'll find items of similar size and shape (coins, pebbles, tokens) grouped together on the foreshore when the tide is out.

From your high point, you'll note how the river surface swirls in a number of different places (Fig. 9). These are eddy currents. They show that something is impeding the natural current, and is changing its speed. So some articles will, at those spots, drop to the river-bed. Mark the eddies precisely, lining them up with

Fig. 9

permanent features on the shore—a shrub, a warehouse door, a bollard. From the water's edge, later, you may not be able to discern the eddies and will need markers, or reference points.

The path a small object takes under water when dropped from a bridge depends on many factors: the point and side of the bridge it fell from; strength of the current; contours of the river-bed, the size and position of obstacles it meets as the current

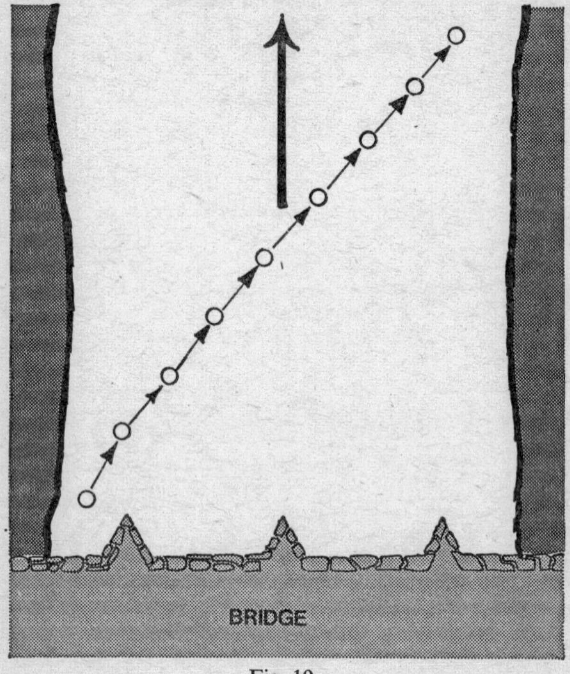

Fig. 10

nudges or tumbles it along. One thing always happens: the dropped article moves diagonally downstream to one bank or the other (Fig. 10). Whether it makes for the right or left bank depends on the dropping point from the bridge and the direction of the current.

If you plan to work a section of river bank repeatedly for a long time, you should toss marked coins or discs into the river from different points on the bridge. It may take a year, or even two, before you come across the first of these pieces. Remembering where and when you dropped them, you'll learn much about the river. Did they carry 100 yards or up to 400 yards, to right or left bank, a long or short way up the foreshore towards the high-water line? Experiments like these help to move treasure-hunting from guess-work towards a reasoned science and a more sustaining pastime.

Also mark in your sketch-pad those high spots that emerge first from beneath the ebbing tide. These are the river's safe deposits: glory-holes. Probe them and you will come across clusters of nails and small bits of non-ferrous metal (washers, brass screws, etc.), all of about the same weight. When the river water was higher, these high spots were bumps in its bed that slowed the current and allowed nails and other articles to sink to the bottom. So nail clusters should beckon you as emphatically as a flashing neon sign spelling out SEARCH HERE.

Preliminary scouting of these areas also trains your eyes to discern different objects. Remember that rings and lockets won't flash and gleam at you as they would from a jeweller's window; buckles won't glisten as they did on the waists and insteps of eighteenth-century dandies. Along with the coins, they will have coatings of mud or moss; silver items will be tarnished black. They will be buried or half-concealed in mud, or lying among the unsightly mess of rocks, broken masonry, rusted metal, gnarled driftwood and sludge. The eyes of Steptoe and Son would light up at the sight, and so should yours. To find the valuables, you must drop on your haunches to make a close-up search, foraging slowly in the debris with your fingers.

On your reconnaissance, you may meet a sturdy fellow in duffle-coat and wellingtons, with a soot-caked pipe jutting from his weathered face. He could be your local man o' the river and, therefore, a mobile reference library. If there's a pub near by, take him for a pint; he'll cheerfully tell you of the changing patterns of river life. His memories of cargo ships and barges, of yesterday's merchants and smugglers, of vanished inns and houses will guide you towards those places where your finds will range from abundant to prolific. That pint will prove a very cheap and worth-while investment.

By now you're a-brim with knowledge and know just where to go to begin your searching. You're buttoned up in old and warm clothing, and have the compleat river searcher's armoury: wellingtons, river rake, sieve, glass-bottomed bucket, magnet, gloves and collecting sack.

The four-inch horseshoe magnet is not vital but is obviously an asset. It's the first item to use in your day's searching with a detector. After you have located your nail patch, run the magnet over the surface of the ground. In five or ten minutes, it will pick up enough surface nails and other iron to ease the day's work. Check this rubbish carefully as the magnet raises it from the mud or shale; you might have picked up old keys, locks, buckles, etc. that are interesting as collector's items in themselves. Put the rubbish in a polythene bag to be disposed of later.

The most practical area to work at one stretch is fifteen feet square, about the size of a living-room. Scratch an outer boundary of this area in the mud, sand or shingle.

Next set up the lines and pins. The first line runs along one boundary of the search area; the second runs parallel to make a yard-wide channel. Begin moving slowly along this channel, sweeping the detector head evenly from side to side.

Make sure the head goes smoothly at right-angles to the lines to cover every square inch (Fig. 11). You'll quickly acquire the knack. Wobbling the detector around in vague, meandering patterns, like a toddler using a paintbrush for the first time, makes a nonsense of having it. The head should never rest on the surface, but an inch above it, a tiny fraction higher if the terrain is very uneven. Beginners often make the mistake of holding the head too high, as if an accidental bump on the ground will damage the works inside the head. The machines are sturdy; they can withstand knocks. So pay special attention to keeping the detector head low the first few times you use it. After a while, you will automatically judge the correct height. Unless the ground is especially rough or treacherous, *always* move backwards, as this instinctively makes you slow your pace. The old fable of the tortoise and the hare could have been specially written for treasure-hunters.

When you've finished working the first channel, move the outer set of line and pins a yard inside the other. If you're moving at a sensible, effective speed, you should take about fifteen minutes to complete one channel without pauses for digging, or seventy-five to eighty minutes to complete the fifteen-foot square search area. When this is done, stake out another. Don't be disheartened if you fail to strike pay dirt at first; the stuff is certainly there somewhere; patience and growing experience will help to find it. Veteran treasure-hunters who can almost smell the valuables in the ground, can have absolutely non-productive days when they are searching new terrain.

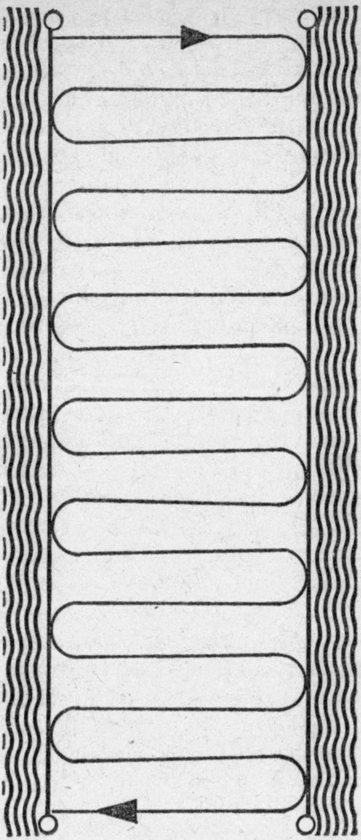

Fig. 11

Truly practised searchers never put down their detectors when working river sites. They rest the head on the damp ground but always keep a grip on the handle, even when trowelling. The unit, which has the brain of the device inside it, must always stay well clear of mud and sand. Suppose you've trowelled up a large blob of mud and the object that caused the warning bleep is still not in sight. You have to pass the lump of mud under the loop, and obviously you will work faster not having to grope around or stretch for the detector handle to pick it up. From the start it's a good idea to learn one-handed digging. This may feel awkward at first, but then so is the business of learning to swing a golf-club with a straight left arm. It should be done for best results. In treasure-hunting you have to stick to the basic rules; it's one activity that has no room at all for unorthodox geniuses.

A superior detector can detect and locate tiny items like rings and coins at a depth of ten to twelve inches in mud, and heavier articles, such as snuff-boxes and cigarette-lighters, at a greater depth.

You don't need to fill in each small hole you have trowelled because the incoming tide will do that for you. However, it's common sense to fill in a deep hole which could form a mudpool and become a hazard. If you return to the site the next day, you could be the poor wretch who steps in the hole and twists an ankle.

The areas of uneven ground where eddies have brought objects are good places to rework by digging alone, especially if they have a lot of iron in them. Bring along your garden sieve, spade and a sturdy garden fork. Squatting on your haunches, thrust in the fork and turn over a small area of ground. About six to eight inches is as far as you need go down. On many sites, you'll hit clay around that depth. Rummage with extreme care, prodding the material with your fork. Methodically work your way down from drier areas to the water's edge. Keep in mind that almost none of the coins or valuables you find will be easily identifiable because of their coating of muck. This is why you must work almost in slow motion. The regular contours of the valuables are the best clues to distinguish them from the rubbish. A treasure-hunter is a kind of detective and, like all sleuths, must train his eyes to search out instinctively the smallest clues—and recognise them. Therefore practice is essential.

Now a few words about using a sieve on the foreshore. In terrain where the mud is loose and oozing, you have little chance of spotting objects. Go to your chosen location as the tide begins to ebb. Stand in the water. Float the weighted sieve and toss several spadefuls of material inside until it is about quarter full. Don't be impatient and heap an immense amount into the sieve; you can't search properly through such a mass. Probe with the trowel, scraping it carefully around the entire surface of the mesh. Even if this yields nothing but pieces of debris, you'll be learning something about the river by noting the size and weight of each piece and where you came across it. It pays to work twice through the material in the sieve before you toss it away. In places where there isn't much metal about, you are better off using the detector rather than the sieve. This is really a tool to use only in areas where your eyes can't discern the surface material or you can't use a fork properly.

Every river, tidal or non-tidal, is a living thing. Waves caused by tides and the wash of passing vessels fret and gouge away at the river-bed and its banks. Several yards below every low-water mark, you'll find a shelf (Fig. 12), either inches or feet deep, that the restless waves have scored out. Dropped articles making their inevitable journey towards one of the river banks can come to rest near these shelves. Currents urge the pieces along the shelves until some obstruction—a log, a mound, a bridge pile—halts them or they glide to rest under a layer of silt, and so are stopped from moving up the shelf. Unless flood-waters or a freak

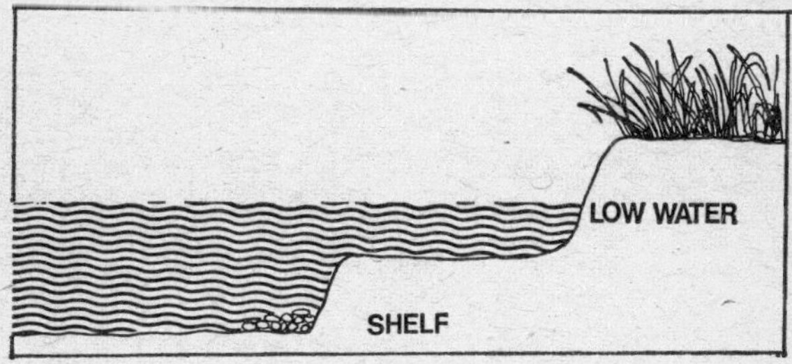

Fig. 12

tide have flung them over the shelf, large objects won't clear the shelf hurdle. So articles made of gold won't make it to the foreshore; they are too heavy. In fact, no more than one article in ten will clear the shelf. Research is an invaluable aid. If newspaper libraries or local people tell of recent floods, you can fruitfully search above the shelf. And it pays to know when spring tides flow in that area; after a spring tide, check above the shelf.

Shelf areas yield the most in non-tidal rivers in exceptional times: dry seasons when the river has shrunk. Otherwise you must work them from the water. In tidal rivers, good finds can come from foreshore areas that contain a mass of junk, from low-tide spots near bridges and in curves of the river.

Rivers are dangerous for the unwary or foolish. You must follow certain basic rules of safety. Before entering any river—no matter how well you think you know it—always probe the river-bed with your rake (Fig. 13). If it drops in a pothole, move a pace or so downstream and test again. When satisfied, step carefully into the water at right-angles to the bank—but *only* at right-

Fig. 13

angles. Later, return to the bank along the same route. *Never* step or attempt to work along a route parallel to the bank; your foot could touch on to very slippery mud or plunge into a hole. You could be in real trouble, especially if you are searching alone, as people do much of the time.

Always divide your search area in the water into lanes, either by line-markers on the bank or visually, and work these lanes only at right-angles. Being alert to every single step you make while in the water is important. At first your concentration must suffer because part of your mind must be conscious of where your feet are and of not making any moves absent-mindedly. After a while, you should have drilled yourself to walk automatically in a river as carefully as you would in a minefield.

Whether your river slips brown and oily through cluttered dockland or flows fresh and lazy through a picture-postcard village, any stretch downstream from a riverside pub is a boon: alcohol has no sense of geography or class distinction. Today's clientele may roar up to the Fourmaster Inn in sports cars to sip gin and tonics elegantly under the mock-Tudor beams of its tap room or on the patio overlooking the river. Yesterday's drinkers, of the eighteenth and nineteenth centuries, were a more raffish and energetic lot. They caroused and roistered through the night; they swarmed to pubs to gamble, to watch cock-fights or bare-knuckle bouts or to dance jigs; they joyfully hurled bottles, tankards and each other into the water. Down the years, all this swilling and jostling dislodged thousands of articles from their hands, throats and pockets.

So, the inside reaches of the first bend downstream from any pub is a fine place to search. If the river curves away from the pub, the best spot would be on the opposite bank. If the down-stream stretch is straight, study the eddy currents, the size of shingle or pebbles on the foreshore. When you've calculated what you think is the most likely place, go on to the foreshore or enter the water there. Work slowly upstream—at right-angles to the bank—using sieve or rake and glass-bottomed bucket.

The current will help you by sweeping stirred-up mud or silt from your footprints behind you. The potential pickings are splendid: flagons, stone ginger-beer bottles, clay pipes, buttons, buckles, half-crowns, brooches, the tosspot squire's watch-chain or tinder box. Look very carefully at all hard lumps in the sieve. Everything you find may look at first glance to be nothing more than a large, hard lump of moss. Some of the pieces will have much more significance than that. So, be alert.

Fords offer marvellous search areas. Pilgrims and foot travel-lers may have stooped there to drink and to sluice away the grime of travel; horsemen have jogged their mounts across or watered them there; iron-wheeled carriages have bounced over the pebbled bottoms, their roof luggage swaying and rocking, the passengers inside jostling around. Highwaymen rated them very highly in-deed as sites to ply their trade. It's easy to imagine a gold watch dropping from trembling fingers or a locket falling from a high-wayman's pocket as he spurred his steed off along the road.

As always, the natural place to search at fords is downstream. But, if the ford has a stone causeway, long-lost pieces may be lodged against it on the upstream side. Use the regular techniques and caution; you don't want to learn the hard way where the ford ends and deeper water begins.

Streams have many attractions for the treasure-hunter. With a waterproof detector, you can search the entire stream bank to bank. Prowl near all bridges, whether ancient rustic or modern horrid; check those verdant banks where families may have picnicked and couples lingered. Streams are less productive than tidal rivers but are a relaxing change of pace and surroundings.

All in all, if you're intent on adorning as fast as possible your sideboard with bottles, buttons and coins, your wife's fingers and throat with jewellery and on recouping your outlays for detector and other equipment, then concentrate on rivers. You can search them every day of the year. And, once you've located and grown to know every yard of your regular river section, it will pay off from here to eternity.

We said earlier that the grottiest and most unappealing sites produce the most valuables. Here are some for Londoners and residents in the south-east to check:

WAPPING OLD STAIRS

This ancient foreshore in the heart of London's dockland is famous in fact and fiction. Sherlock Holmes and Dr Watson trod its slimy reaches with their lanterns glowing dimly in the fog; Edgar Wallace's Lascar seamen and Chinese opium runners drifted silently along it; Dr Fu Manchu got up to nasty business here. In real life it has bustled with life for centuries. Merchants, boat builders, urchins, dockers, sailors reeling to and from their ships have inevitably lost an enormous amount of stuff. Try coinshooting there among nail clusters at low tide.

BLACKFRIARS BRIDGE

This is near the site of a famous recent archaeological find: a 500-year-old wooden boat. You would applaud treasure-hunting as a fine hobby if you did nothing more than work the 200-yard stretch running westward past Festival Hall on the south bank. Massive rebuilding throughout the 1960s helped to erode the foreshore considerably, making it more productive than ever. With diligence and luck, you could have enormous success here.

VAUXHALL BRIDGE—LAMBETH BRIDGE

The Tate Gallery stands midway between the bridges on the northern side but you won't find any gallery discards in the mud. What you will find are astonishing assortments of coins from any country you care to name. It is also most promising for English and foreign artifacts lost there during the past 250 years. This stretch is a brilliant example of the boon that urban stretches of the Thames are for treasure-hunters with keen eyes and a disregard for muddy clothing.

GREENWICH

This long, twisting stretch of the southern Thames foreshore, from a point just east of the Royal Naval College to the Woolwich dockyards, is inexhaustible. Pennies, halfpennies and shillings are among a multitude of coins, dating back to the fourteenth century, which have been discovered there. It also contains an abundance of nineteenth- and twentieth-century coins. For obvious security reasons, you are not allowed near the foreshore at Woolwich Arsenal.

FULHAM

The section of the north bank from Chelsea to Battersea Bridge will keep any searcher active. It yields a plethora of coins from the reign of George II onwards, rings and many military objects: Victorian bullets, buttons and badges. The width of land between the river wall and the low-tide mark can offer up coins and artifacts from many ages. Passing craft frequently scrape their keels on the river bottom. They dislodge from their sanctuary in the mud dropped objects which fetch up on the higher reaches of the foreshore.

LONDON BRIDGE TO SOUTHWARK BRIDGE

A most exclusive area: along the south bank is very good hunting-ground for Roman coins, and virtually nothing else. The existing London Bridge is sited just upstream from the original Roman London Bridge. The foreshore on the exact site of the original bridge has been dredged, disturbing its ancient caches; bronze Roman coins can be picked up by using eyes-only method. A good place is the ridge which is revealed at low tide. This is not a good area to use a standard pulse-induction detector.

In the north of England and in Scotland, river stretches in the industrial areas of large cities and towns (Sutherland, Stockton-on-Tees, Aberdeen) can yield marvellous dividends. They, like the stretch of river in the Ruyton district of Newcastle, are the equal of the Thames for finds. Any place where ships, fishing vessels, or any type of commercial craft have plied will have its repositories of coins and artifacts; many spots will be glory-holes. Search near bridges and, as usual, concentrate on nail clusters for potential finds.

FORTH BRIDGE

Both north and south sides of the Firth are good; a mind-boggling amount of coinage has been tossed from the bridge down the years. The most productive section lies between the Binns and Hopetoun Head. On the northern side, search about forty to fifty yards downstream from the bridge. The south bank brings a larger harvest; some treasure-hunters have literally shovelled out coins, ranging from Victorian to Elizabeth II.

BEAULIEU FIRTH at INVERNESS

Stand in the river and work either side downstream from a foot-bridge that spans the Firth here. You will find a few pre-Victorian coins and a plentitude of others of more modern reigns.

CHAPTER 8

BEACHCOMBING: HOW TO LOCATE AND PROSPECT ON THE RIGHT BEACHES

ANYONE who hunts constantly for lost objects on a beach is called a beachcomber. The image of beachcombers in the public mind is very fuzzy, having been shaped largely by reading popular fiction or lurid Sunday newspapers. From fiction comes the stock figure: the unshaven derelict in grimy rags. He peers through a gin-haze from beneath his battered straw hat as he lurches along a golden Tahitian beach with outrigger canoes ploughing through the surf in the background.

He is a joke figure among the local white residents, who are unaware of his secret: that he fled from civilisation (which was naturally within a square mile of Grosvenor Square) to protect the family honour or cover up for a fellow-officer who had embezzled the mess funds. The impact of strong sun and raw liquor has swiftly reduced him to a piece of human flotsam.

In many people's minds the domestic beachcomber also has some sordid aspect about him. They visualise him as the lone, shabby figure seen poking around in the sand in the evenings by couples taking a post-prandial stroll to escape the penitentiary atmosphere of their boarding-house. With some distaste, they imagine he also rummages in the litter-bins along the esplanade.

In real life, which is usually far removed from half-baked fancy, the contemporary beachcomber is a much different sort of person. He is a vacationing public schoolboy with tailored corduroy jeans tucked inside his wellingtons as he digs away at the sands of Aberporth Beach on Caernarvon Bay. He is the bank securities officer, with some historical flair, from his near-by week-end cottage looking, with sublime optimism, along Charmouth beach in Devon for relics of the famous ninth-century battle there between local inhabitants and Danish pirates. (Whatever ancient pieces he finds he'll hand in to the local Archaeological Society, or send to the British Museum.)

He could be a member of the House of Lords in blazer and flannels crouched over his detector as he searches for the heirloom bracelet his wife thought she lost while playing badminton on the sedate Middleton-on-Sea beach in Sussex. (One of the most sophisticated peers, Lord Clark, renowned for his television series *Civilisation*, has a detector but is not the model for the gent mentioned above.)

Nowadays beachcombers abound because the pickings are so good. Beaches rank second only to tidal rivers for the sheer volume of stuff that the searcher can find. Well over half a million articles are lost on British beaches each year. And who can wonder at this? The way people disport themselves at the seaside makes it an accomplishment *not* to lose something.

Coins, cigarette lighters, pipes, removed wristwatches, rings and bracelets can, and do, tumble from your pockets as you sit or sprawl in the sand; as you carry bundles of badly folded clothing; as you loll in a deckchair; as you play beach football or cricket; as you do handstands to impress the blonde in the red bikini; as you dash to the water's edge to rescue Junior in the shallows. Something may be lost when that wretched black dog knocks over your picnic basket. The ring you forgot to take off may slip from your finger which has contracted in the water. The clasp of a pendant may be broken by the slapping of the waves or the rigorous movements of the jousting game you've been playing in the water. One treasure-hunter commented, with watering mouth, that during a hot summer the mass of lost objects under a popular beach forms what seems like an almost unbroken carpet concealed beneath the sand.

Even when owners realise on the spot they've lost something, the chances are it will stay lost. Dry sand is voracious in gobbling up dropped articles. If you toss a coin or small disc (such as a brooch) into the sand on edge, it vanishes immediately. If it falls flat, a small spurt of sand from a passing footstep or a puff of wind covers it all too effectively.

An immediate search with eyes only is an ordeal: one patch of sand looks depressingly like any other. Although myriad stretches of Britain's 6,000 miles of coastline are searched by amateur eyes-only beachcombers, ranging from the hawk-eyed to the inept, they overlook an astounding amount of lost articles.

So the harvest is there for the shrewd and methodical searcher: he can't really fail. As with all types of prospecting, the keywords are preparation and patience.

Obviously you must know as much as you can about tide times, tidal streams (currents to you and me), prevailing winds, and the places that people seek out on and near beaches. All such data tells you where things are most likely to be dropped and where they're most likely to be moved by the restless water.

Tide movements around Britain's coastline are so consistent they can be predicted with only a few minutes' variation for years ahead. Every day we get two flood (incoming) tides and two ebb tides. It takes the water about six hours and twelve minutes to flow one way—high tide to low, or low to high) and twelve hours and twenty-five minutes for the full cycle: high to low to high. This means that if full tide is noon on Monday, it will reach its height again between 12.30 a.m. and 1 p.m. on Tuesday.

Each year port authorities publish tables showing times of tides in and out plus the size of the rise and fall of water in Britain's major ports. These tables also have sets of figures called constants that you add to or subtract from the main area times to work out tidal movements at any point along the coast. Many local newspapers in seaside resorts carry tide tables for their areas. You can also learn them by asking fishermen, whose awareness of the sea's dangers makes them most obliging to polite inquiries, or near-by Coast Guard stations, if you are planning to search in a new area.

Fishermen are the best sources to tap. They'll also tell you of winds, the time of year to expect storms, of shipwrecks, ancient smugglers—almost the whole gamut of things you should know.

Twice a month, during full and new moons, we get what are called spring tides, when tides rise and fall through their widest range and currents surge at their strongest. Naturally when spring tides are running, the agitated water tears ruthlessly at the beaches and treasure-hunting is enhanced. It is also handy to know that currents run their swiftest during mid-tide, between the third and fourth hours since the tide turned. Neap tides, when tidal movements are least, come during the moon's first and last quarters.

Depending on the character of the place you visit, there are plenty of visual signs marking high-tide points. Black slime builds up on jetty and pier pilings to the high-tide level, as it does on a sloping line along beach groynes running down into the sea. On the sand itself, you can easily make out two different high-tide lines. The one farthest up the beach, marked by pieces of driftwood, shows how far up waves have pounded during storms. The lower has lines and clusters of pebbles that tell where the latest tides have exhausted themselves.

The treasure-hunter working an isolated beach without clear man-made landmarks (jetties, boat-sheds, near-by houses, etc.) must know about currents in the area. The local meteorological office can give some information. This will be limited—currents are so unpredictable—but is a sound basis to work on. Again, the best people to ask about currents are local fishermen. From knowledge handed down, perhaps through generations in one fishing family, they will gladly tell you what they know. Where are the currents strongest and where weakest? What unseen obstructions can cause counter-currents? Where have landslips occurred and is the rate of erosion high? (The majority of older coins are found under heavily eroded beaches.)

One of the most pertinent facts that fishermen will pass on is that currents are like a shrew's temper: tricky, dangerous and ever-changing. Even the simplest currents change direction four times a day with tidal movements. But a headland, breakwater or other obstacle can increase a current's speed up to five times. A restricted harbour mouth, islands, etc. set up fierce counter-currents. Eroded beaches have been gnawed by strong tides and currents—threshing water has moved and deposited many objects—so they are good places to work with your detector. On all isolated beaches you'll learn much about currents and possible good prospecting stretches by noting where the flotsam and jetsam have drifted.

Your searching will be easier when you've seen both the lighter summer tides and stronger winter tides worrying at the same beach. Summer tides build up the beaches; winter tides wear them away. Most importantly, you'll note the movement and force of water needed to produce valuables and just where they are likely to be.

Above all, implant this piece of information in your brain: it is the ebb tide which has the force that brings coins and other lost articles to the surface. The crashing waves of the incoming tide swirl and loosen the sand or shingle. These persistent incoming waves—or swash—slam downwards and are too heavy to do more than shake the sand and disturb whatever is lying there.

You can gauge the movement by tossing a coin into the backwash. Instead of slithering into the sea with the receding water, the coin will jitter its way up the beach. It's a peculiar sight.* And it sets up a mystery that we'll consider in detail later.

You have to realise that each of the hundreds of British beaches is unique. No other can be subject to exactly the same intricate pattern of factors: location, prevailing winds, main alternative winds, curve and width of bays, direction and speed of currents, the whirl and tug of counter-currents, the mass and shape of

* You can only see this during the first hour of ebb tide. After that the ebbing tide does not have the force to move coins up the beach.

spits, number and position of man-made features, plus the promontories, inlets, composition of beaches, rate of erosion, and so on. Knowing from where and how hard the prevailing winds blow is not enough by itself: local winds can be mostly westerly but strong undercurrents may pull articles on the sea-bed in the opposite direction.

In general, currents in British coastal waters follow this pattern of movements, as do objects dropped into them: south coast—west to east; east coast—north to south; west coast—south to north. Local residents will fill you in on variations.

All these points underline that you can't afford to be casual and half-baked in your research.

The three big questions to answer for profitable beachcombing are: where? when? and how? People first began to frequent British beaches in large numbers around the middle of the nineteenth century, the newly-laid railways playing an important role. Then bank holidays, which were initiated in 1871, established the habit of the seaside outing and spurred the growth of the country's seaside resorts.

Of course the style of holiday clothing, bathing costumes, the range of facilities have all changed enormously since then. Basic patterns of behaviour have not. People still do much the same things on beaches—sunbathing, picnicking, frolicking, strolling, courting—now as they did a century ago. So they still go instinctively to the same areas of beaches.

It's a good start to get hold of old picture postcards of any popular beaches you intend to work. These are still to be found in seaside second-hand bookshops, curio and antique shops, hundreds of cards depicting the beaches as they used to be in years gone by. They'll show you where the piers, bathing machines, fun fairs were in Victorian and Edwardian times. The kids taken to Southend, Brighton or Blackpool to celebrate Queen Victoria's Jubilee romped in just those areas where today's children build their sand castles and play beach cricket and leap-frog. Where the children are, the parents are also, losing articles in their tens of thousands.

So the really sensible treasure-hunter goes *where* the people go on resort beaches. But not *when* they are there in their thousands, running, jumping or staying still. Summertime is not really the best period of the year for searching. For one thing, tides are feebler in summer months; quieter water does not disturb lost objects so much, guiding or hurling them closer to the surface of the sand. For another thing, you can't search effectively or courteously when the sand is almost invisible under a layer of prostrate pink-and-white bodies. Daylight hours are long in summer, but prospecting hours on popular beaches are not. In midsummer you can start at daybreak and coinshoot above the high-tide mark. Don't bother much about the lower stretches between the high- and low-tide lines: they are not very productive in these months. You won't have long hours to yourself. Holiday-makers begin to drift down from their hotels, boarding houses and caravans to the beaches around 7 a.m. By half past eight even a long, wide beach can be crowded. People who have come there to relax don't want you nosing among them with your detector. Besides, you can miss so much ground that this kind of futile and irritating treasure-hunting is a waste of time. Do the public, other treasure-hunters and yourself (in that order) a favour by eschewing it when beaches are crowded.

Go down with your detector when the crowds have gone. (A sharp shower will empty a packed beach in late afternoon in just a few minutes. If this should happen at, say, 6 p.m., you would still have some three hours of searching in midsummer.) Then the work is easy: lost pieces will be only a few inches beneath the surface. The main handicap will be the profusion of junk (bottle-tops, silver paper, etc.) that will excite your detector and irritate you. Like commuting in packed, sweaty and cruelly uncomfortable trains, being exasperated by rubbish is a nuisance you'll get used to enduring.

Summer prospecting above the high-tide line is the only time when you're likely to find rings in the same general area as coins. Once the strong autumn tides and storms have done their jobs, coins and rings will separate. Coins are lighter and tides can push them surprising distances very quickly after they are dropped. Two things prevent rings from travelling as far. If gold, they are much heavier than modern coins. And the hole in the centre (in treasure-hunting terms it's called a trap loop) gives the ring more pulling power against currents to lie close to where it was dropped. Silver rings, being lighter, are carried further and can sometimes turn up among coin clusters out of season.

The best times for summer prospecting are immediately after storms. Heavier seas will have moved the stuff around, so you can fruitfully search beach slopes between high- and low-tide lines.

Oddly enough, your most profitable summertime prospecting could be at night with nothing more than a hand torch and a collecting bag. In many parts of Britain, local seaside authorities send their earth-moving excavators and bulldozers on to beaches to dig up sewage pipes and let effluent flow into the sea. It's a common operation and can take place nightly in the one locality, although most daytime visitors never know that it happens.

These huge machines uproot tons of sand, biting holes eight or nine feet deep that their blades fill up again soon afterwards. This gouging out of so much sand and shingle flings last month's lost objects to the surface. You can pick items off the freshly-revealed top layer without a detector. Local residents will know about this work; they get used to the growl of the machines at night. A polite inquiry to the council engineers or sewage officers could be your means of finding out where forthcoming excavation will be taking place. In some areas, notably the West Country beaches, you could go on fourteen nocturnal excursion's during a fortnight's holiday.

For safety and good manners, don't blunder in the way of the workmen or their machines. Getting accidentally buried under nine feet of damp sand may be inexpensive but it's an ignoble way to go.

Despite these benefits, summertime is best used for observation. Learn where people gather, then work those places in the out-of-season months. Let's quickly consider an immensely popular beach like Great Yarmouth, perched on the rump of the Norfolk coast. What are potentially the best prospecting sites along its five-mile stretch of sand?

First, the piers. There are three: Wellington Pier, The Jetty and Britannia Pier. All are splendid points, as are jetties, groynes and all man-made structures on beaches. Fishermen and sunset-gazers lean over pier balustrades losing things from their hands and pockets. Strollers and scampering children drop items that slide through gaps between shrunken boards. If currents and tides are right (and research will show if they are), piers are splendid coin traps. Around Britain's shoreline a startling number of coins in particular are lying underneath or alongside piers.

Two of Great Yarmouth's summer shows are held on Britannia and Wellington Piers. So tens of thousands of vacationers tread along them each summer, blithely adding to the treasure-hunter's potential finds.

In the season you can devote several days to digging and detecting beneath and near the piers. You'll be overwhelmed by the volume of your finds.

As you scan the packed beaches, note the location of beach-huts, toilets, deck-chairs, tents, windbreaks, steps leading to the sands from the esplanade or roadway above. See where people embark and disembark from pleasure-boats and launches; where the hot dog, ice cream, soft drink stands and refreshment kiosks are; where the children's ponies trot; where wives drag their reluctant husbands to watch the bathing-beauty contests. Study and jot down all places where people assemble and where finds are most likely to be (marked by crosses in Fig. 14). Those are the spots to work, either in the quiet summer hours or, preferably, in the off-season.

Fig. 14

Groynes are those timber-frame, low broad walls that thrust into the sea to check the drifting of the beach and the encroachment of the sea. Where you find them, you'll see the sand has piled higher on one side. The high tide has nudged the sand and other things along to build up there. People tend to sit on the higher side. So work it rather than the lower and more eroded side.

On isolated beaches without obvious man-made landmarks, you'll have to pay greater attention to prevailing winds. People sit or lie in the lee of these winds, sheltering behind hummocks of sand, bluffs or whatever is available. Secluded corners of bays are promising places for treasure-hunting.

However, spits of land running into the water are not really worth while. Many people do sunbathe or pledge their eternal (i.e. to the end of the vacation) devotion on them. But currents around spits carry dropped articles away almost as soon as they fall. They are swept where you can't get at them.

Pay special heed to areas showing strips of coin-sized shingle, especially where these cross the high-tide line. It's here the current has become too weak to haul lost coins and other items any further. These are glory-holes that may contain large clusters of coins and small valuables of the same size. Other possible glory-holes are the shallow pools you'll see glinting here and there on the slope between the high- and low-tide lines when the sea is right out. Currents have very likely carried and concentrated coins into these depressions in the sand. They bear close detecting and digging. To strike a glory-hole can be very exciting. It seems to justify those long hours of research, the days of lean pickings, the aching cold and lashing spray of the grimmest expeditions.

So much for the 'where' of beachcombing. Now let's talk about the 'when?'. We've already suggested you grab your detector to work between the high- and low-tide lines after summer storms that have fretted the sea bed and tossed up whatever is lying there.

For bigger hauls and the older, more valuable coins, wait until the season is over; it usually tails off at the end of September. The best time for beachcombing is October. It brings some of the heaviest tides that gnaw into sand and shingle. Check your tide tables and try to search when the fiercest tide of the year is running. Also keep your ears alert for radio weather-forecasts or ring the meteorological stations. As soon as you hear that the first autumn storm is on its way, organise yourself so that you can be ready to work your checked-out beach the moment the storm has blown itself out.

You'll have to wrap up well against cold winds: wear anorak, or pac-a-mac, sweaters, gloves and wellingtons.

Take your detector, line and pins, plastic sieve, trowel and trenching tool. To take only a trowel as you would for searching commons, etc. is wrong: you may have to dig a hole three or four feet deep. We'll go into techniques in a moment.

From the end of October and through November your finds will become fewer and fewer no matter where you are operating. From December onwards, go to beaches only on gale days. As before, you must be poised for action to get cracking on the beach the instant the wind's fury has died and seas have calmed. During these winter months, you'll have to rely solely on storms to smash sand and shingle apart and disgorge their caches. High-tide lines are the first places to look.

In the wake of a terrible storm, say, directly after Christmas, you may find artifacts that have lain under the sand of beach or sea bed for two years or more. The overwhelming proportion of articles of any kind that you'll discover on British beaches will have been lost there the previous summer, or the one before that. It is becoming rarer and rarer to find on a beach one of those later George VI or Queen Elizabeth II pennies (which an American writer once described as the only pennies he had ever seen that you could kill a person with) which officially went out of circulation when decimal currency was introduced early in 1971. You will certainly find no old coins in your summer prospecting. Only big storms unearth the older stuff. It's not known whether it is flung up from the sea bed beyond the low-tide line or from beneath the beach sand.

Contrast this with river prospecting. Given the same amount of searching as a beach, there is not a single river in the country that won't produce coins that date back 500 to 600 years. And the usual range of beach finds is far more predictable: mostly modern coins, cigarette lighters, rings, keys, toys, cutlery, pendants, some bottles and not much more.

This is the mystery we touched on earlier. Man has lost things on beaches as long as this land has been inhabited. Since bathing became a recreation of the people in the nineteenth century, losses have been mind-boggling. Literally tens of millions of old coins, from farthings to sovereigns, plus mountains of artifacts have been lost down all those years. The search for them has been incessant: a multiplying army of beachcombers has been at work using more and more sophisticated equipment.

Yet the overwhelming bulk of these riches have not been found. They are not necessarily under the beaches, but they are certainly not far away. But where? They have vanished. It doesn't matter if you have 20/20 vision or the finest detector in existence, nor does it matter how experienced you are or how scientifically you forage. You won't find this stuff.

Some treasure-hunters have dug immense holes, many feet deep, on deserted summer beaches (and, we add, filled them in again) but they have still not produced old coins. Unless you work in the immediate wake of a storm, you very, very rarely dig down to a level that delivers Victorian, or even Edwardian, coinage. The tides carry almost all dropped articles away from the beach and deposit them in some unknown place, below the low-tide line. How far is anybody's guess. This is why people who dig for worms on the flatlands of beaches as the tide recedes find few lost articles.

The distance of an object's journey to its unknown destination must vary according to the strength and range of tides, the pull and thrust of currents and counter-currents, the physical contours of the sea bed, the nature of obstructions. Remembering how currents work, dropping objects to the bottom in relation to their weight, there is an unbelievable hoard of coins and valuables somewhere beneath the restless waves. When this tantalising secret of the sea is revealed, it will yield an absolute bonanza on every beach. The glory-hole to end all glory-holes. And there are hundreds and hundreds of them around the British coastline.

The puzzle has remained unsolved for a simple reason: people have not investigated it deeply enough. Sophisticated treasure-hunting, that dates from the early 1950s when light detectors were innovated, is still in its infancy. The army of amateur treasure-hunters has had what in the span of history is a mere blink of an eyelid to investigate our hidden riches. Archaeologists and professional treasure-hunters are few. So we still know very little about the scope and profusion of what lies under our soil and the waters of our rivers and seashores. It could take several generations before we have a really clear idea of what is down there to be found.

One very experienced treasure-hunter has been pondering this mystery for years. Bit by bit he has been unravelling it; he believes he is very close to solving it. When he does, he visualises his private pilot flying him back regularly from his villa in the south of France in his private helicopter to continue his British treasure-hunting with a solid gold detector.

But back to reality and the 'how?' of the techniques of good beachcombing. Most of the time treasure-hunting is a solitary task; treasure-hunters seem to prefer it that way, free from distractions of other people's movements and conversation. Like sitting in a condemned cell, searching focuses the concentration marvellously. And you must keep your whole mind on the job; you're looking for a great many items that measure less than one inch across.

Still, the first search of autumn after a storm can be just the time for a team—four to six are a good number—to work effectively. If the just-departed summer has been fine, the team can work the entire high-tide line—it may be one long glory-hole.

After a wet and unsettled summer, the heavier-than-usual seas will have already washed lost items down from the high-tide line. So start some four or five yards down-beach from there.

If you decide to start as high up the beach as you can, mark out the boundaries for the day's work by scratching lines in the sand or shingle. Allot each member a fifteen-foot square to work, moving down-beach towards the low-tide line in a series of paths parallel to the water. They can detect, trowel down to eighteen inches, then dig down to two or three feet if finds seem to warrant it. If the first October storm has sent you rushing to the beach very soon—say a week—after the bathing season had ended, you may not need to go lower than eighteen inches for excellent hauls.

The solo treasure-hunter can improve his financial standing with tide-line coinshooting along both high and low lines. This is easier but far less lucrative than systematic searching with line and pins. Centre on the tide line itself. Walking backwards, follow it while gliding the detector head slowly in straight paths at right-angles to the line. Make each sweep six feet, i.e. three feet to either side of the tide line.

Bear in mind what we've said repeatedly about patience. If you're a newcomer to treasure-hunting, you may not have chosen the spots as well as you could have done. So your first day could bring disappointing numbers of finds. It's most unlikely you'll be as lucky as the television reporter who found a gold ring within two minutes while cameras were recording him searching a common. No matter how regular you are at saying your prayers, don't expect providence to be sitting on your shoulder.

You could take days, perhaps weeks, before your selected beach begins to pay off handsomely. But don't be disheartened. Remember that nature doesn't change. If your research is sound enough, and your patience firm enough, the beach will offer up its hidden valuables. And once you've cracked the code, that beach is yours for life.

Paste in your hat the simple information that the high- and low-tide lines and the terrain between them will produce your best finds (Fig. 13).

The bane of prospecting popular beaches is the enormous amount of metal detritus including beer cans, soft-drink cans and hairpins you'll come across. It doesn't matter what kind of detector you use, you'll be plagued by some items of this excess of ferrous and non-ferrous metal rubbish. With a pulse-induction unit, you'll find a seemingly endless collection of hairpins; with induction balance, or BFO, an equally aggravating bounty of silver paper. You'll just have to grin and bear it. For beach-combing, the old dictum of northern industrialists and mine owners rings true: where there's muck, there's brass. On a very popular beach full of debris, a standard garden-rake can clear away much of the surface junk before you bring your detector into play. Where the volume of rubbish is less, you can just as easily dig it up.

Always carry a container to put the rubbish in. A shoulder-slung canvas bag with several pockets is best: rubbish into the larger pockets, valuables into the smaller. We don't expect you to be idiotic enough to sling silver paper and ring pulls out of your way. You'll either find the same items as you move further along the beach or the next tide will wash them back into the same place. Also, local authorities appreciate having less refuse to clear away and there will be a smaller amount of dangerous stuff, like sharp-edged ring pulls, to slice open the hands and feet of next year's bathers.

By using common sense, you will be consolidating the right and correcting the wrong image of treasure-hunters; showing they are not litter-louts or despoilers of the areas they search.

After working the tide lines, make a more general search, on a part of the beach you've already charted as likely (near the bathing cabins, refreshment stands, steps, where the currents could have carried pieces, etc.). Clear the surface rubbish. Mark your fifteen-foot-square patch and lay down the lines and pins. Work backwards in a path parallel to the water line. Keep the detector head low because the coins and jewellery could have sunk deep.

If one spot offers up more modern coins than others, you've struck a likely glory-hole. This is a spot for careful digging (down two or three feet) and using your plastic sieve with the detector. Where you find a cluster of coins, it's simplest to dig down the first eight to twelve inches with a trowel. Don't fill the sieve: about fifteen trowelfuls at a time will do. If the sand is damp (and it certainly will be below eighteen inches) your arms will quickly ache from the weight of the loaded sieve. With too much sand in the sieve, there's also the danger of overlooking items lodged in corners where the mesh meets the sides.

Pass the detector head over the load. If you get a reading, sieve slowly and carefully. If not, toss the load away and refill. Many coins sink below the range of the best detectors, so digging and sieving is the best method for achieving results. Work this way as well in the deeply banked sand you always find on one side of groynes and breakwaters.

After you've exhausted finds at the shallow level, dig down to about three feet. This naturally is slower because you'll be shovelling and probing in wet sand. But even one hour's searching at the right spot on a beach can bring you within easy range of the fifty coins that veterans say should be the minimum day's haul for coinshooting.

When you've finished for the day and intend to come back, mark the exact spot, preferably with two reference points (level with the third post of the beach groyne to the west and opposite the door of the boat shed to the north), and record them in your notebook. A convivial evening, worry over your tax, love problems or the passage of time could wipe them from your memory.

An area above the high-tide line that can pay off is near a sea wall. People lose stuff when sitting on, leaning against or jumping from sea walls. Most are reinforced with iron, so it's pointless to approach closer than two or three feet with a detector. If the area is producing well, then dig right up to the wall. In dry sand, sieving work will be fast.

You have read the laws of treasure-trove for beach finds earlier in the book. I will confine comment here to a few brief points.

On quiet public beaches, authorities can no more crack down on a treasure-hunter roving with a detector and digging than they could arrest a toddler for building sand castles. What you do with your finds does, of course, interest the authorities. Items found above the high-tide line could be classified as treasure-trove, depending on the material they are made of (silver or gold), below that line they become the property of the Receiver of Wrecks. The simplest thing would be to take whatever old coins or articles you find, no matter from which part of the beach, to the nearest police station. When you've specified exactly where you located it, the police will know what to do.

Perhaps it would help if existing laws were recast, stipulating that finders must hand in articles above a certain established value and all objects they consider of potential historical interest. But that is purely academic. You may discover something—perhaps a snuff-box or old turnip watch—so tarnished by oxidisation that it must have lain buried for sixty or seventy years. The owner is almost certainly dead, but by law you must hand the item to the authorities. Obey the law implicitly; you won't lose by it in any way; you'll be almost certain to reclaim the article after the twelve-week period it must rest with the authorities has elapsed. One treasure-hunter has handed in thousands of pieces down the years; not a single one has failed to come back to him. Always list every single item in your own notebook before handing anything to the police.

To cater for the super-righteous, we might as well state the achingly obvious point that it is a waste of time handing in modern coinage to the police. Most likely the people who dropped the coins hadn't even realised the fact. And the police won't relish the needless paper-work of having polythene bags of sand-encrusted coins cluttering their already overcrowded offices. They'll be very tempted to drop the bundle casually on to your toes, especially if you are wearing thong sandals.

CHAPTER 9

A SELECTION OF BRITISH BEACHES: WHERE AND WHEN TO SEARCH THEM

IF you work diligently gathering information, then searching sensibly, you will quickly be recouping, from beaches near where you live, your first outlays on equipment.

Working different sorts of terrain, pursuing different types of finds, will make you keen to travel and work others. Here are a few of the hundreds of British beaches you can profitably prospcct.

ABERYSTWYTH

The handsome and sedate Victorian and Edwardian buildings that look down on this popular resort beach, make some holiday-makers feel they should be wearing neck-to-knee bathing costumes. However, its interest for treasure-hunting is most certainly contemporary; it is a place to look for modern coins. Areas near the pier should be the most productive. Check the busy places—where deck-chairs are set out, children's play section, high- and low-tide lines in the 'safe bathing' area between the markers—to work during quiet summer periods: early morning, late evening. Note that undercurrents are strong near Constitution Hill, on the northern part of the beach.

BEACHY HEAD

The southernmost point of eastern England, the great grass-topped cliff is reached by the Downs Road. At night the 142-foot Royal Sovereign lighthouse rakes its beam down into the Channel from atop the 534-foot-high white cliffs. Not long ago two hoards of Roman coins were found at the cliff top. Now more coins can be found on the beach, possibly from other hoards that have tumbled down with eroded chunks of the cliff face. Search both tide lines thoroughly. Offer your Roman finds to the local archaeological society; you'll have a plentitude of modern coins to keep for yourself. Beachy Head has a Lover's Leap, so keep alert if you're prospecting there during the phase of the full moon.

BLACKPOOL

A short visit to Britain's largest and most famous resort could bring you enough coins to pay for your own holiday—in the West Indies. For best pickings, follow the tide out in the early morning before the tower organ starts booming. Blackpool has been a popular watering-place since the eighteenth century, so you can come across Victorian finds. Blackpool may be the discarded silver-paper capital of the world, but ignore this nuisance as you prowl both high- and low-tide lines reaping your splendid harvest of coins.

BOGNOR REGIS

Queen Victoria's affection for 'dear little Bognor' seemingly lured many of her loyal subjects to frequent it. This would explain the rich lode of Victorian coins (plenty of modern as well) to be found between Bognor and Bognor Rocks at the western end. Search the entire beach very conscientiously: detecting, sieving, digging—the works. With gold prices ever rising, this is one beach that might make you rich, if you're industrious and patient. During the 1960s, a gold bar was found on this beach. The British Museum assay showed a platinum content, an irresistible clue that it came from South America. It's not too wildly romantic to suspect that the sea and beach hold more gold bars.

BOURNEMOUTH

If coinshooters do go to heaven, they will find the surroundings familiar: they will look very much like the long, sandy beaches of Poole Bay, in Hampshire—especially in the Bournemouth section. The superabundance of coins is nothing less than you would expect to find beneath the sands of one of Britain's most patronised and best-equipped resorts. In fact, Bournemouth is so fecund for coinshooting that the law of diminishing returns may have set in. The theme song of Bournemouth treasure-hunters could be 'You'll Never Walk Alone'. During the summer months especially, people with metal detectors often seem to outnumber the bathers and shapely European girls there to study English at the numerous colleges. If you are gregarious and enjoy being part of a squadron of treasure-hunters by all means try your luck here. If your view is that treasure-hunting is most satisfying as a solitary pursuit, you would be better advised to seek less frenetic spots. Shrewdly chosen parts of Scotland and Wales would be ultimately more rewarding for you. If you decide to work Bournemouth beaches, go to them only when they are deserted, not when they are teeming with bathers or any other holidaymakers.

BRACKLESHAM BAY

On this part of the six-mile sand and shingle-backed beach that runs between East Wittering and Selsey Bill in Sussex, you'll see fossils on the rocks when they are uncovered at low tide, and should have good finds of Victorian and modern coins. The beach is restricted at high tide. Roman coins and trinkets have been discovered here. An intriguing sixteenth-century cannon, found a few years ago, was a strong pointer to an unknown wreck and possible rich finds.

BRIGHTON

If you decided for some strange reason to work only one British beach, Brighton must be a candidate for that exclusive honour. It will take you further towards putting a deposit on one of its elegant houses than almost any other beach. All those well-heeled London commuters, the retired brigadiers, day-trippers in 'Kiss-Me-Quick' hats, baffled Continental tourists and ubiquitous film crews lose vast amounts of coins and valuables.

In winter, when the madding throng has dispersed, work high- and low-tide lines, but be warned that waves can be dangerous at high tide in stormy weather. Wait until the seas have subsided completely. It is a lovely beach to search in the dawn hours of summer. But it's after gales you can find eighteenth-century silver and copper coins, plus gold coins, and enough silver paper and modern junk to send your detector beserk.

CAISTER-ON-SEA to GREAT YARMOUTH

This stretch, running north to south, is almost as far east as you can go and remain on British soil. Although the area is soaked in history (Caister Castle dates back to 1432), almost every coin you'll pluck from the sand will be modern. Yarmouth's North Beach yields somewhat better than other stretches, but pickings are well sprinkled along the entire stretch. It's unprotected from the sea's full assault, so waves can froth and boom far up the beach; sieving work above the tide line could be most productive.

CAMBER

Camber Castle lies to the west of Rye harbour and is one of a number of artillery forts that Henry VIII built (it's shaped like a Tudor rose). It was originally on the coastline, now it's a mile inland. The land between the castle and existing shoreline is mostly silted terrain, ending in a narrow beach.

This silted area is strictly for the ultra-patient treasure-hunter who is prepared to make many fruitless visits there. He'll be looking for very rare old coins that are buried in small numbers. Because it's non-tidal, you can work the area at any time. Muster the concentration of a brain surgeon while making your intensive line-and-pins searches. A single find here will be an enormous thrill and, ultimately, we hope, very lucrative.

CARDIGAN BAY

This presents a good area to visit for modern coins. There are several places along its reaches well worth searching, especially after early autumn storms. The best of these is Abersoch, a lively resort on the western arm of the bay, with a long sandy beach. Other spots to try are Criccieth, which has a ruined castle on sentinel duty above it, and Morfa Bychan, with a sandy beach known as Black Rock Sands. Avoid the south-eastern end; it is unsafe for bathing because of powerful currents, so finds would be few.

CHESIL BEACH

Chesil Beach is an eighteen-mile-long natural breakwater of shingle that runs north-west along the eastern end of Lyme Bay from Portland towards Abbotsbury in Dorset. It is a fine beach for team digging that can be so tough it will separate the dogged from the timid. So will weather conditions. The best times to search are in the December–March period after the savage storms that lash the breakwater, which itself was formed in one storm of demonic fury. What you'll need for Chesil are strong backs, sou'westers, oilskins and rum tots against the chill.

Nature has worked in your service. Throughout thousands of years, tides and storms have tossed and tumbled millions of pebbles until they have been sorted most precisely into sizes. At the Portland end they average three and a half inches across and shrink gradually to around one inch at Abbotsbury. It is towards this latter end of the beach that you should coinshoot with your machine, in those immediate post-storm times when the riches will have been lifted within detector range. Search very slowly; many coins are tar-covered or badly corroded and thus hard to identify at first.

Scattered on the sea bed within the embrace of the arc of coastline between Start Bay and Portland Bill are twenty-two known wrecks, ten of them unidentified. The four closest to Chesil Beach are one unknown ship and three submarines. From time to time, silver coins and bars are washed on to the beach, presumbably not from the submarines, unless the crew of the German UB74, sunk in 1918, were surprisingly well paid.

CROMER TO EAST RUNTON GAP

Running towards the northern end of the hump that is the Norfolk coastline, this beach faces squarely to the North Sea. There is plenty of sound advice on winds, currents, etc. to be gained from local fishermen, as well as spooky tales about the near-by sandbank known as 'Devil's Throat'. The beach, its sand dotted here and there with shingle patches, has many shallow pools revealed at low tide—thus lots of glory-holes. If you work systematically you can find coins at all levels of the beach.

One interesting facet: Cromer really became fashionable in the nineteenth century, yet you can find quite a few 200-year-old coins here.

DYMCHURCH

This resort, nestling behind a grassy sea-wall and dating back to Roman times, has a sandy beach backed by a strip of shingle. Inland lie the Romney Marshes. Along the beach, which ranges from gravel to fine sand, you can find coins from any period of the last 300 years. Your best aids here are detector and trenching tool. Work the low-tide line. If you come upon a cluster of coins, dig down eighteen inches to two feet. If you're lucky and diligent, you don't need to find many coins to make the sweat and effort worth while.

FLAMBOROUGH HEAD

Like an outer layer of loose, coarse skin, a reef runs around this nose of headland jutting out from the Yorkshire coast. The beach lies at the foot of 150-foot-high grassy cliffs on which is perched the Flamborough lighthouse. Tides are strong and dangerous. Head here after storms and search very carefully at the absolute low tide, when you'll see many rock pools. The gold coins that turn up here regularly could be washing in from an unknown wreck.

FOLKESTONE

In summer, bathers throng East Cliff, where the firm sandy beach, pleasantly free from any shingle, is a mass of deck-chairs, windbreaks and beach huts. So hauls of modern coins are good. The big lure for treasure-hunters are Elizabeth I florins that are thought to have come down from the cliff-face through erosion. Big seas prise them up from the sand. Come here after storms or spring tides to work the low-tide line in particular. These florins are worth upwards of £20 each in reasonable condition, surely lure enough to focus your concentration, make you ignore the aches of digging and come back many times, even if first visits are barren of finds.

HAYLING ISLAND

This vaguely kangaroo-shaped island squats in the large indentation on the Hampshire coast, between Portsmouth to its west and Thorney Island to the east. South Hayling has a popular resort beach of quietly sloping sand and shingle that runs for six miles along the entire length of the island's southern coast. The narrower eastern end, towards Eastoke Point, where currents are swift, has a number of groynes that should be carefully searched.

We've been told a genuine treasure is buried somewhere in the beach; old coins have been discovered in the Hayling Bay area, which is midway along the beach and where shingle is heaviest. This treasure could be a companion box to the brass one containing seventy-seven coins of the sixteenth century that was dug up in the mid-1960s. This box was said to have been hurled back into the sea after a bitter legal wrangle over its ownership.

Divers in the area have retrieved several cannon with different dates, strong clues to at least two different unknown wrecks. You'll get best hauls of modern coins following both tide lines. Seas can be strong, so also check the ground above the high-tide line where you are able to get at it. Altogether, a beach to work and rework many times.

HELMSDALE, SUTHERLAND AND WICK

We have grouped these two areas together because both are so far to the north, treasure-hunters travelling from distant places should consider trying to work them both on the same trip. Helmsdale, the more southerly, is a tiny lobster-fishing port snuggled between cliffs at the mouth of the Strath of Kildonan. A supper of lobster—they are among the most succulent in Europe—will help to thaw you out after a day's arduous prospecting.

You'll need some kind of reviver because this beach and Wick are two of the coldest spots for out-of-season prospecting in Britain. Only the most dedicated, responsible and ambitious searchers will want to, and really should, visit them. Playboys stick to Brighton. Pack with you knitted caps, oilskins, sturdy gloves, several of your thickest sweaters and your heaviest trousers (you may need to wear two pairs to prevent your legs from freezing).

You can't evade the icy cold because autumn and winter are the times to go there, after vicious North Sea gales have battered the coastline. A prodigious force of wind and seething water is needed to produce what you'll be mainly seeking: relics of Viking plunder. Along the Helmsdale coast, Viking money-horns have been washed ashore. The mouth of the river, close to the rocks, is the place to search and dig.

December to March is undoubtedly the best period for Wick, and tough work because the area is exposed to Arctic blasts. You could seek weather information from the divisional rescue headquarters of the Coast Guard (Wick 2332) situated on the South Head of Wick Bay. Scan, very, very carefully, near the rocks of Wick Bay and work the small inlets. Wick is a resort with two open-air salt-water pools, so modern finds should come your way as well. Since only tough, no-nonsense people will go here, they'll enjoy the rugged charm of the two villages and the straightforward pithiness of the fishermen and villagers.

ISLE OF MAN

If you visit the capital, Douglas, for the motor-cycle racing (Tourist Trophy in June, Manx Grand Prix in September), you could take a detector to search the wide curving beach of Douglas Bay. In June, you will have to restrict yourself to dawn and late evening patrols; you won't see the sand for the layer of reclining flesh during the day.

Hoardmaster John Webb tells me four hoards were found on the island in the last century. Coins bearing dates from the eleventh century to the fifteenth century still turn up on some beaches. Your best bet is the resort of Ramsey in the north-east, which has a long, curving beach of sand and shingle. Work the shingle first, then the sand with line and pins if you have time. The half-mile-long Queen's Pier area could produce enough modern coins to pay for your return air-ticket to the Isle of Man.

On the south-west coast, seek the old coins at Castletown, whose ancient fortress, Castle Rushen, is wonderfully preserved; it's in far better repair than many of the bathers. If tides are right, you could try the sand and shingle beaches on the long, skinny toe of land that is Langness Peninsula on the western sweep of Castletown Bay.

LIZARD POINT

The first clue to Lizard Point, the southernmost point of England, is that it's classic smugglers' territory. In fact, a pirate built the first lighthouse here, supposedly so that its light would lure ships to founder close to land he owned there. For centuries, this thrusting, wave-hammered mass has been a landfall for ships. So it's not surprising gold coins and pieces of gold bars have been washed up along its awesome coastline. Here the cliffs rise up to 180 feet above the waters of the Atlantic. You can pick your path down the cliff face to the beaches, but it's exhausting and dangerous; the timid should stay at home and work the local common.

Confine your searching between Polpear Cove and Pentreath Beach. The exertions of descending the cliffs may leave you so breathless that you have to work very slowly, which is to your advantage. Scrupulously check every tiny cove and inlet that offers some shelter, however slight, from the surging ocean. This too is a place for repeated visits, as are all beaches with older coins and valuables concealed under them.

KINGHORN, FIFESHIRE

As treasure-hunting grows as a hobby, more and more people will seek out the still neglected beaches of Scotland. Naturally, you won't expect to coinshoot with the prolific success you'll achieve at teeming southern and other resorts. Two things offer much delight on Scottish beaches: the superb scenery of many and the valuable older coins they contain.

Kinghorn, a sandy beach with cliffs behind it, stands on the northern side of the Firth of Forth and is a good sample of the charm of Scottish beaches and the good prospecting they offer. Scottish noblemen have been coming here to restore their jaded tissues since King David I proclaimed it a Royal Burgh in the twelfth century; hence the ancient coins. Then perhaps travel the three miles east from the small Kinghorn beach to Burntisland, which offers a small sandy beach near the docks and a long grey stretch that is almost a mile wide at low tide. The usual resort structures and study of the currents will guide you to the most likely prospecting spots.

MARGATE

In those summer months when London's streets are thronged with tourists from all over the world, look for native Londoners in Margate, their main seaside haven for 200 years. With the interplay of so much sand and humanity, you can coinshoot abundantly in summer during those early and late hours when you won't get your detector wrapped around your neck by irate bathers.

To find anything older than Queen Elizabeth II decimal coins, come here swaddled heavily in those December–March days immediately after gales have whipped along the South Channel. Two hundred years ago this area teemed with more smugglers than you'll find bikini-clad nymphets there today. Their pastimes were rowing booty to and from France and 'paultring', which in local dialect meant wrecking.

Perhaps aided by wreckers, the East Indiaman *Hindostan* sank near by in 1803. The wreck is almost certainly yielding up some of the old coins you'll find by detecting and digging the storm-ravaged sand. Have a good detector and follow the tide out. Finds of late seventeenth-century coins around here point strongly to another unrecorded wreck lying close by. Energetic and studious detecting after a Force 9 gale could send you clinking happily back home. If you find anything that should be reported to the Receiver of Wrecks, do hand it in. You've a good chance of ultimately being awarded it.

MOELFRE, ANGLESEY

On the night of 25/26 October 1859, the most horrific recorded storm in British history blasted and bellowed along the Welsh coast, wrecking 114 ships in that single night. The frenzied north-easterly hurricane swept on to rocks, then broke the back of the screw-steamer *Royal Charter*. A total of 459 of her crew and passengers were drowned. To the bottom with them went an estimated £700,000–£800,000 in gold bullion being transported back from the Australian gold-fields. More than half of the bullion was later recovered; divers have tried unsuccessfully to salvage the rest. The devastating impact of the storm, plus more than a century of unceasing tides and currents have spread the remaining raw gold and coins far and wide.

Moelfre's shingle beach, part protected by a small headland, is not large. Potential rewards are so brilliant that it would be worth taking a late (October) holiday to spend several days working as much of the beach as you can. Start by detecting close to the cliffs. Whenever your detector registers, dig well down at that spot.

MONEY HEAD

The Rhins, the slim hammer-head of rocky land that dangles awkwardly from the southernmost point of Scotland's west coast, is another area only for stalwart treasure-hunters. Chilling south-west winds slash across the land; the air is alive with spray and old legends. Some well-polished tales of treasure reputedly buried in these parts would beggar the imagination of a Robert Louis Stevenson. At least one treasure ship did go down: the sea has fetched up old coins in widely scattered parts of Money Head between Ardwell Bay and Cairngarrock Bay. Search carefully the rocky areas, among outcrops and on the shingle.

PORTLEDGE

Some of the most vivid chapters in England's nautical history have been shaped by men raised along the deep and far-stretching curve of Bideford Bay on Devon's north coast. A few years before the Spanish Armada, Queen Elizabeth I learned that one in six of her sailors were Devon men, a large number from this area. The awesome wrath of the Atlantic—it pounds with a force of 2,000 lb. per sq. ft. on the rocks at Hartland Point—has spelt even more tragedy for Spanish seamen than the cannons of Drake's fleet. For instance, the original timbers of what is now the Portledge Manor Hotel—which owns the small private beach accessible through a break in the cliff-line—came from four wrecked Spanish galleons.

These waters abound in wrecks, many of them unknown, that have surely been the source of many articles washed on to the shores of Bideford Bay. Where there is access you can search close to the water's edge at low tides; but be very careful the tide does not steal up on you as that could be very dangerous. Seek permission at the Portledge Manor Hotel, which is a short way inland, before going to the beach.

While in this part of Devon, you could head north to coinshoot on the broad swathe of surf-pounded beach at Westward Ho, which is backed by a ridge of grey pebbles that the sea diminishes by one yard each year.

PRESTATYN AND RHYL

If many communities in North Wales have a deserted, ghost-town look in summer, it's because most of their citizens have bowled along the A548 and A525 to these energetic resorts. Both have a plenitude of attractions that make them first-class coinshooting beaches. Prestatyn squats between its wide, sandy beach and green wooded hills that rise sharply behind it, and is a profusion of holiday camps, golf-courses and other places of recreation. Get hold of a detailed local map and plan your best searching areas, which lie between the camps and golf links. Prestatyn's role as a lead-mining centre in the nineteenth century means there are plenty of Victorian coins to be found (most likely after rough weather) as well as thousands of coins that today's affluent society unwittingly leave behind.

Rhyl has the entire gamut of seaside facilities. In both resorts, you'll do best working the tide lines when the beaches are quiet. Rhyl has fast and tricky currents flowing at ebb tide, so be careful where you stray; keep especially alert at low tide.

SELSEY BILL

Swirling currents have chewed away acres and acres of land from this headland since the Saxons made their first-ever landing on English soil here in the fifth century. Erosion continues unabated; no British beach has been more diminished in the last hundred years. The strong force of water constantly hauls Victorian coins and valuables up from places that the sea has claimed. Local life-boatmen or H.M. Coast Guard (Selsey 2274 or Shoreham 2226) will give you the low-down on currents and tidal movements. When you have studied these, search the most exposed areas of the shingle beach, and accessible places close to the high sea-wall that protects the east beach.

WATCHET HARBOUR

Only a contemporary lyric-poet who was a connoisseur of mud would be inspired to dash off a sonnet to the beauties of this commercial port on the Bristol Channel. (Although Samuel Taylor Coleridge wrote 'The Rime of the Ancient Mariner' after gathering information from an old seadog here.)

West of Watchet the Channel waters are deeper and so the shore is freer of the unsightly mud, but in the port itself the coasters loom up at tilted angles in the slimy mud at low tide. Wear wellingtons and take a waterproof detector for your searching, and dig the small beach carefully. Your best finds will be eighteenth-century coins that crop up here regularly and in splendid condition.

CHAPTER 10

BOTTLE-COLLECTING

In recent years, growing numbers of Britons have taken to the bottle. Or, more accurately, the bottles, for bottle-collecting has proved to be one of the sturdiest and most durable offshoots of treasure-hunting.

Certainly a large number of imposing bottle-collections had been gathered together before treasure-hunting became a widespread hobby; the charm and fascination of bottles have been evident for centuries. Also, you don't need a detector to locate bottle dumps likely to contain collectable specimens. In dumps where tin-cans, springs and other metal objects are found, the glassware will almost certainly be modern bottles and jars like those you drop nightly into the trash can in your kitchen.

Treasure-hunting has converted thousands of people into ardent bottle hunters through stimulating a general interest in all manner of collecting. Even materialistic types, who took up the hobby originally for the money they could make from their finds, discovered they couldn't resist the pleasure of starting some kind of collection. Hence the burgeoning arrays of coins, buckles, musket balls, medallions on mantels and desk tops and the specially-built display cases that grace many a living-room corner.

Bottles quickly won themselves a multitude of admirers. The sites where they are buried can be relatively easy to seek and locate. On the most productive sites, perhaps where nineteenth-century breweries and glassworks once stood, they can be dug up by the score. The seemingly infinite variety of shapes and colours and sizes allows the collector to quickly amass rows and clusters of dramatically contrasting pieces.

With bottles, the specialist instinct comes swiftly into play. Before he knows it, the individual has advanced from general seeker to bottle-collector to addict. His craving focuses on Codd's stopper bottles, or Warner's quack cure bottles, or aristocratic cobalt blue castor-oil or poison bottles, or ginger-beer flagons, or very old Hamilton bottles with pointed bottoms.

At a time when prices of antiques have climbed faster than anything in the American space programme ever did, bottle-collecting has been one of the least expensive ways of gathering Victoriana. And collectors have the assurance that every item they have dug up personally, or bought, is genuine. No furtive industry has sprung up, or will spring up, turning out phoney Cherry Tooth Paste pot lids or imitation Bellamine Flemish jugs.

Above all, bottles are brilliant proof of Keats's famous line: a thing of beauty is a joy forever. Even the lowliest of them, such as the small, plain-glass inkwells produced in their tens of thousands in the late nineteenth century, delight the eye. They also project an intriguing sense of history. Glance at one of those inkwells and you can visualise a Dickensian clerk or hapless schoolboy scratching away with his quill by flickering candlelight on a late November afternoon.

You can acquire the same sense of history from gazing at other, more exotic bottles. For instance, the Eclipse poison bottle, which dates from the mid-1870s, has an exaggeratedly nipped-in waist that helped its owner to distinguish it in the dark. Its silhouette revives memories of those Victorian and Edwardian ladies who squashed their insides beneath whalebone corsets in their quest for the fashionable eighteen-inch waist.

Antique bottles of almost every kind convey, along with the nostalgia, a sense of style to the beholder. Very few of them were without some kind of distinctive feature: the maker's name or an advertising slogan in embossed or incised lettering; crests, scrolls or labels permanently stamped into their sides; fluting; screw-in stoppers; flared, sheared or moulded lips and so on. Some were asymmetrical or the glass was spattered with air bubbles so that each bottle seemed unique, even when it came, in fact, from an early production-line.

Like most other products, bottles have been both easier and faster to make in this century through more efficient processes. Yet smoother production-lines have not been an absolute blessing: they have reduced the character and distinctiveness of containers. Foods, liquids, beauty products, medicines that all once came in handsome glass bottles are now packaged in plastic bottles and cartons or treated cardboard boxes. The housewife who has tottered back from the supermarket with two, or three, laden carrier-bags bulging with provisions gets no uplift from surveying the myriad containers they hold. Milk bottles, beer bottles, coffee jars, cheese-spread cartons, cold cream jars, aspirin and cough-mixture bottles—what a dreary and mundane line-up they are. Even the perfume bottle she brought back from her last holiday abroad, or received for Christmas, may be functional and uninspired. So it's no wonder that the appealing bottles of yester-year have become so popular.

Many splendid bottles belong to the recent past. Ginger beer, for instance, was sold in stone bottles and flagons until the 1930s. They can still be found in abundance throughout the country, so rarity does not explain their immense popularity. The answer may be simply that they are very good to look at; their squat, glazed silhouettes and the profusion of stamped labels they bear make them interesting to array along shelves and window ledges. The jumbo sizes, which hold two gallons, are more than just ornaments. Home brewers, disgusted by the high prices and taste of modern beer, sometimes use them for storing their powerful concoctions.

In the early 1970s, bottle-collecting spread swiftly on two fronts, among treasure-hunters and purchasers. When the demand asserted itself, specialist shops, selling only antique bottles, were established in a number of cities and towns. (It led me to set aside one entire section of my original metal-detector centre in Biggin Hill as a bottle emporium.)

For treasure-hunters, finding bottles is not a long or complicated job. Bottles either ended up in general rubbish dumps, special bottle dumps or under or near bodies of water—rivers, ponds, lakes and streams. Wherever you live in Britain, you should not have to travel far to find them.

Not surprisingly, the history of bottle dumps runs almost parallel with the history of the Industrial Revolution. Bottle making on a relatively large scale gathered momentum in the mid-eighteenth century, pre-dating the stirrings of the Revolution by a couple of generations. First came large numbers of wine bottles, then, in the early nineteenth century, beer bottles. Lacking today's range of materials (plastics, metal cans, aerosol dispensers), the Victorians put everything, from colognes to tea-leaves, into glass. It was their all-purpose material, along with earthenware vessels (which are still widely used, notably by the Dutch, for liqueurs and apéritifs).

As industry expanded in the last century, more and more glassworks were established to meet the demand for containers. Many communities that were little more than villages had at least one, and often more, glassworks. Most of these have long since vanished, but the dumps into which they put their rejected or outmoded bottles may still exist. As well as these 'pure bottle' sites, there are bottles to be found wherever rubbish was dumped: official and unofficial tips, quarries, disused mines, wells and land raised up or levelled with refuse so that it could be built upon.

Having read previous chapters diligently, you'll know where to seek information to lead you to these sites. As usual, your local library is a good place to begin researching. There you will find, or can order through the national lending service, histories of the Industrial Revolution, books and almanacs describing the growth of the area that interests you. Maps are also helpful, either perused at your library, or with town records and planning registers, in your municipal offices.

News columns in old newspaper and magazine files will tell when official dumps were established and went into disuse; of court stories detailing complaints and prosecutions for illicit dumping. Also study advertisements in these publications and old catalogues. They will show the bottles used for alcoholic and other drinks, medicines, unguents, lotions and quack remedies guaranteed to cure everything from falling hair to bunions. The advertisements will feature the names and addresses of manufacturers, agents and perhaps wholesale and retail outlets; all these locations could be the sites of dumps or forgotten storehouses. One firm of bottle sellers in London obtained much of its original stock through locating a 'frustrated' shipping order of eighteenth- and nineteenth-century bottles.

All the above sources will also give you leads (locations of former industries near river and canal banks) about likely bottle sites close to water. Obvious places to go are river fords, the premises of old or former riverside pubs or downstream sections from them, and river bends where slower water will have allowed bottles to drop to the bottom. Don't forget the bridges. Drinkers inflamed by their contents have either flung or sleepily dropped satisfyingly empty bottles from them. Schoolboys of many generations have liked to hurl bottles as far as they could upstream from a bridge, then shied stones at them as the current drifts them along. Periods of strongly contrasting weather—dry seasons or immediately after strong storms—are good times to seek bottles. (You should search for them in the same way as for rivers, outlined in Chapter Seven.)

Of the small armoury of tools needed for bottle hunting, you should already have a glass-bottomed bucket and folding shovel. A digging fork you can buy at any hardware store. You can make a metal probe rod by welding a foot-long piece of steel rod to a four-foot length of rod, forming a T-shape. Either affix rubber handles to the short horizontal piece or wrap tape around it at either end. You will need a secure grip because the probe is thrust into the soil above a prospective site and turned clockwise until the tip clunks against bottles—a sound your ear will quickly identify. When you have located a dump, you should probe at three-feet intervals until you have plotted its exact boundaries. By doing this, you'll avoid shovelling or forking dirt from one part of the dump on to another and creating needless work for yourself. Retrieval can be summed up in one word: dig.

To carry your finds, you could use a large canvas hold-all or large plastic bags or, if you can bring your car fairly close to the dump, bottle crates that are partitioned into compartments. Take along plenty of newspaper to wrap around each bottle for protection, and polythene sheets to separate the layers, if you think this is necessary.

As with all types of prospecting, find out beforehand who owns or has authority over the land you wish to enter. Then you will know whether to approach a farmer, property company, river board, port authority or whatever. Declare your objective fully, stress that you know your responsibilities and will clear up whatever you have to disturb; you should find that most of them are obliging.

A few words on cleaning. Good bottles will have lain under soil or mud for years. They will have achieved an equilibrium down there and must be given time to adapt to changed temperatures and humidity once they are retrieved. So let them rest for twenty-four hours before cleaning.

Unless you keep coal in your bath or actually bathe in it, you'll need a special tub—perhaps a disused baby's bath, stone wash-tubs in your old laundry, a cut-down 44-gallon drum, or inflatable wading pool. Whatever you use, half fill it with cold water, then pour in and stir a pound of washing soda. You'll have two buckets or pans near by, one filled with plain water, the other filled with the water in which you dissolved a pound of soda. Rinse the bottles one by one, first in the plain water, then in the soda bucket. During the second rinse, fill the bottle to the top, cover its mouth with thumb or forefinger—to keep air out—and lower it into the bath. Keep topping the soda bucket until you've put all the bottles in the bath, where they will stay for either two days (those relatively clean to begin with) or a week for really grubby ones.

Rinse the adequately cleaned bottles under running water, then either dry them out of doors or rub them dry with a cloth.

Those still bearing interior stains should have a handful of fine gravel, then a little water, poured into them. Cover the mouth and shake the bottle for, say, two minutes. Then wash with clean water. Outside stains can be treated by rotating the bottle in a bucket filled with fine builder's sand. Weak acids can remove deep stains from bottles but beware of using them on stoneware; a thorough internal scrub should suffice for stoneware.

Enthusiasm for bottles covers as wide an age span as any segment of the collecting habit. At the upper end, nostalgia is the key. The sight of a creamy-brown Miller and Childs ginger-beer bottle can set a grandfather to remembering the bottlefuls he swigged at dormitory 'beanos'. Fondling the lid of a Rowland's Otto of Rose cold-cream jar, a silver-haired matron recalls, perhaps with a blush and a sigh, that it was the first beauty aid she dabbed on her adolescent cheeks. Both grandfather and matron are now too old to endure the rigour of searching and digging up their own bottles. They are fortunate that the supply of bottles and growing number of outlets can allow them to indulge their collecting fancy in a sedentary way.

Delight in bottles can readily encompass all types, from sheared lip inkwells to cut-glass cologne bottles with hinged and chased silver clasps and tiny flush-fit stoppers. Yet many people find their major interest veers without prodding towards making them either predominantly glass fanciers or stoneware fanciers. For the glass addicts, happiness is garnering bell-shaped inkwells, Georgian wine bottles, Lipton's tea bottles, a nineteenth-century beer bottle with a Henry Barrett internal screw stopper, a dozen long-necked cobalt blue hair-restorer bottles, a glass screw feeding-bottle with tubing and teat still in place and a Clarke's Blood Mixture bottle ('The Mixture . . . never fails to cure Scrofula, Scurvy, Scrofulous Sores, Glandular Swellings and Sores, Cancerous Ulcers, Bad Legs, Secondary Symptoms, Syphilis, Piles, Rheumatism, Gout, Dropsy, Black Heads or Pimples on the Face, Sore Eyes, Eruptions of the Skin and Blood, and Skin Diseases of every description').

The stoneware fancier doesn't need Clarke's mixture to improve his blood circulation. His therapy comes from finding or buying two-tone (they are the most prized) ginger-beer bottles of all sizes, a flask-shaped Adaptable Hot Water Bottle & Bed Warmer, and a flock of elegant pot lids—Burgess's Genuine Anchovy Paste, Holloway's Ointment, Woods Areca-Nut Tooth Paste and, the most coveted of all, Cherry Tooth Paste—Patronised by the Queen. (Colour Plate No. 7 shows a selection of pot lids.) One of the most enthusiastic pot-lid collectors in Britain is the accomplished comedian Leslie Crowther.

Bottles can simultaneously stir the senses and line the pockets, which is a very pleasant fusion of minor art and finance.

CHAPTER 11

PORTRAITS OF THREE ENGLISH TREASURE-HUNTERS

TREASURE-HUNTERS come in all shapes and sizes. Some cycle to their favourite sites with their detectors slung at their shoulders like carbines. Others cruise there in the family's third car, a shooting brake, with a bottle of Chablis in a portable fridge to sip for 'morning tea'. Some stalk mucky river sites peering beneath the same flat caps they wore when exercising their previous fancy of pigeon keeping. Headgear for others are the golf hats with ornately embossed crests they have worn while gouging divots on the courses at Nassau, Los Angeles and Manila.

The ranks of people absorbed in finding and/or collecting range from 'roll-their-own' miners to Princess Margaret, who collects and tumble-polishes semi-precious stones. From 1973 onwards, the pastime gathered very strong momentum among the prudent middle classes. This jumble of solicitors, butchers, bricklayers, accountants, together with judges, and entertainers like Leslie Crowther and Jimmy Young, have applied a patina of respectability to the activity. The casual observer has become more reluctant to paste the label of 'crank' instinctively on to the figure with the metal detector. He knows that person could turn out to be his MP, a renowned film director or the man who will have the final say on his application for a bank loan.

The common link among all these people is not, as some critics would have it, the greed for riches; many of them are already too well-heeled to bother about the extra money their prospecting could bring them. The spark is the one thing without which any work or pastime is basically meaningless: enthusiasm. With prospecting, this enthusiasm can advance swiftly and decisively into a kind of passion. This deepening of interest doesn't depend on the treasure-hunter unearthing a constant bonanza of golden valuables; some people would be content retrieving from soil and water nothing more than pot-lids or buttons. It comes from the satisfaction of achieving some success through hard work—the long slog of intelligent research, enduring the chill winds of tidal estuary or winter beach and the muscle-straining effort of locating and retrieving objects. In one word: dedication.

Those who have already done some treasure-hunting will have learned all this for themselves. So criticisms of their motives and habits will have assumed their proper perspectives. For newcomers, and apprentices as well, I want to try to convey that dedication in this chapter with tales of three treasure-hunters. Two of them are professionals and the third an amateur perhaps best described as a disciplined fanatic.

First the professionals. The more experienced of the two is the long, raw-boned man whose name crops up often in this book and whose bearded face you'll see in a number of the photographs. John Webb. John has been a professional since 1958. Much of the money he earns to sustain himself, his wife and three children comes from selling the objects he finds. Almost certainly, John has recovered more articles from under British soil and salt and sea water than any other individual. Each year, he scoops up an average of 30,000 ancient and modern coins; he has tracked down more hoards than anyone else in this country; he is the man who is confident he will solve the riddle of the sands (see Chapter Eight), the ultimate destination of the unaccounted-for mountains of beach losses.

Here is a partial list of the things John Webb has found, and which any prospector has the potential find: to axe-heads, badges, bicycle lamps, bracelets, brooches, bullets, buttons (military and civilian), cameos, cane ferrules, cannon balls, candlesticks, car horns, carriage lamps, chain purses, charms, clay pipes by the score, cigarette lighters, coins, cuff-links, cutlery, doorknobs, firebacks, flat-irons, flintlocks, fob seals, gold teeth, Bren guns, hair slides, hatpins, hoes, horse brasses, hunting horns, insurance plates, jugs, knives, lead seals, lockets.

Other items are match strikers, medals, medallions, musket balls, napkin rings, necklaces, pens, ploughshares, pendants, rings, shears, snuff-boxes, toy soldiers, metal spectacles, statues, taps, telescopes, thimbles, metal toothpicks, vases, watches, watch charms and whistles.

John Webb set out the basic design for one of the range of detectors I sell, the Coinshooter. He has worked hard to establish clubs to link treasure-hunters together and has made repeated overtures to archaeologists, seeking to reduce the antipathy between them and treasure-hunters. He cheerfully submits to the physical anguish that only a dedicated man will endure pursuing his interests. He has prospected chest deep in mud, on the beds of rivers where the slime made it impossible to see even his hand before him. John has slept under canvas while prospecting Wick beach in the dead of a bitter Scottish winter so intense that the weight of hoar frost threatened to collapse his tent.

John accepts these discomforts, not with the resignation of the punch-drunk commuter, but with satisfaction. He is one of that tiny minority in the population: the man who has deliberately chosen and pursued his true calling. As a tow-headed boy of seven in London's East End, he decided that he would be a professional 'searcher-for-things'; the phrases 'treasure-hunter' or 'prospector' had not yet entered his vocabulary. He did seven years apprenticeship as a tiler but, as soon as he qualified, cast aside security to chase his dream.

His neighbourhood, as a lad in Stratford, was a fine proving ground, scarred with buildings that first the Luftwaffe, then later the wreckers' ball had torn apart. He worked them all over from loft to cellar, ripping up floorboards, tapping for hidden cavities in fireplaces and staircases, rummaging in gardens and along the river banks. It paid off; he supplemented his pocket-money selling the old coins, rings, brooches, etc. that he found. He delighted in his vocation. On the contrary, a schoolmaster was taken aback when John told him his ambition was to be a professional scavenger.

Soon after he took up the work full-time, John spent four years in a row gold prospecting alone in a remote and beautiful part of the Scottish Highlands. On the first two trips, pickings were thin; on the last two he averaged 22 ounces of alluvial gold each visit; it was worth £13 an ounce. Now that gold prices have climbed into the stratosphere, he plans regular visits to the area when other commitments will allow him an extended stay.

On one of his first contracted jobs, John took a Geiger counter on to the Essex marshes to check some strange and immense stone buttons that frightened workmen had excavated. Each button measured three feet across, and looked like a grinding wheel with a hole in the centre. Some justified the workmen's fears by giving a Geiger-counter reading; some didn't. John believes the objects were buried there somehow early in World War II, when most people would have thought 'radio-activity' meant putting together an ITMA show. After John had checked them fifteen years later, they were taken away to be processed and made harmless.

Down the years, many people have asked John to search their homes to look for family hoards, family heirlooms, jewellery, weapons, even the individual ring that a distraught housewife had lost. The three times he was called in to locate hoards proved so much wasted effort: the people were either backing a vague hunch that something was around or they were poorly fed with their information. John is much cannier now about these requests. 'Other times it can be marvellous to see delight and surprise light up in the faces of people when you find an item, especially the small thing of sentimental value,' he says.

His personal searches for hoards have been much more successful. He has located two highwaymen's hoards, one of them consisting of thousands of George I and George II coins.

I'll recount in detail how John found part of a miser's hoard. This should serve two purposes: revealing how a professional goes about the business of prospecting and underlining that there are no short cuts to finding hoards.

John got his first clue when browsing through a late eighteenth-century newspaper in a London newspaper library. He read a brief paragraph about a man called Andrew L . . ., a recluse who had just died in Hertfordshire, apparently a wealthy man. This suggested the tantalising equation: recluse + wealth = hoard. So John checked other newspapers in the British Museum and Guildhall in London. He uncovered a few leads but these petered out because the records that might have helped were among those lost in a bombing raid during the last war.

So he went to Hertfordshire to study the registers of a parish church. Sure enough, there was a death entry. As was the custom of the day, the local clergyman had written a few candid words alongside the entry. (This delightful custom is no more, and it seems a pity. If the deceased had been a popular, and presumably God-fearing man he was ushered off to eternity with glowing words. If his church-going was sparse and his contributions to the church even sparser, he was dismissed with a bad notice.) Andrew L. earned a few terse and quickly-scrawled words: 'he died as he lived, in poverty.' This did not tally with the original newspaper story that he had been a man of substance.

From local newspaper files, John learned he had lived in an old manor house on some sixteen acres of land that fronted a back road between the old village of Baldock and Letchworth. The house was not marked on nineteenth-century ordnance maps, and local records gave no information that hinted at wealth. However registers in a neighbouring parish said he had farmed about 150 acres elsewhere, a good holding for those days.

John pored once more over local records and, during intermittent research over the next nine months went through almost fifty books relating to the area to see what more he could discover. The sum total of information was not much: the man had been known as a local recluse, he had lived frugally, and the sale of his effects after death had raised very little money. This strengthened the suspicion he had concealed some, if not most, of his wealth. His contemporaries might have been curious about him, but one book dealing with local characters had warned people to keep away from him.

John did locate his house marked on very old ordnance maps, but knew these can be notoriously inaccurate in some regions. He double-checked the location against the later, and better, maps and was relieved to note that the boundaries of fields, hedgerows and small roads had changed very little. After receiving permission from the owner of the land to look around, he came across pieces of tile and brickwork that marked where the miser's house had once stood. This took him about fifteen minutes. He estimated the boundaries of the land that had belonged to him, and decided he would have to search an area of two acres.

Now he had to gamble his time, effort and money that demolition crews who had pulled down the old house some time in the 1840s had not found the hoards themselves. He decided they had not, and this was more than a casual guess. In earlier centuries, the regulations on treasure-trove were very strict. It was a criminal offence not to disclose it to the authorities, i.e. hand it over to the landowner. If workmen had found a cache and not reported it, the landowner would have found out somehow. Money has to be spent and word would have filtered back to him from local tavern-keepers, merchants and villagers. So, one way or another, it was almost certain to have been recorded somewhere. Intensive research had brought John Webb no mention of such a hoard.

John now got permission to dig from the current landowner, who jokingly told him he was wasting his time and that he could keep everything he found. John put down his line and pins and set to work. He had to travel a round trip of seventy miles daily from his Essex home to the site. For seven days straight he toiled (a higher work rate than is recorded in Genesis) and found nothing.

After an hour's work on the eighth morning, his detector gave a moderate buzz. He dug down eighteen inches and the blade of his trowel struck something hard—a pewter pot.

'I almost fainted with jubilation. I had uncovered a hoard of silver and copper coins from the reign of George I—halfpennies, pennies, shillings and a few crowns. A total of 283 coins.

'With quaking knees, spilling and retrieving coins, I took it back to show the landowner. He shook my hand, wished me good luck and said all he asked was that I kept absolutely quiet about it; he didn't want people swarming all over his fields.'

John made a careful inventory and took the pot to the nearest police station, an automatic reflex with him and, we hope, with all other treasure-hunters. He realised this hoard could not have been all that the miser had accumulated; there should have been at least another six pots—or any other kind of container—each holding as much as the first.

Pressure of other work prevented him from returning to the site for another four months, by which time the landowner had sold out and was in the process of moving. After six more fruitless days, John's quest was abruptly halted. The new landowner came down, inquired what he was doing, listened sympathetically then told John he really didn't want him digging on his land. So that was that and the remnants, indeed the bulk, of the hoard still lie there undisturbed as the whole lot had done for two centuries.

John later recovered the pewter pot and coins from the police. He sold the stuff in bits and pieces over the next year, making several hundred pounds. Not a fortune for all the time and sweat he had put into finding it. He felt he might have made more money by selling it as one lot and recommends to other treasure-hunters that they explore this arrangement before disposing of any hoards they find. In retrospect, the thrill and excitement of locating this hoard was very much tempered with regret that he had to wait for so long before returning to the site; that delay very likely cost him a lot of money.

For hundreds, perhaps thousands, of contemporary treasure-hunters, John is the Pied Piper who lured them into taking up the hobby. They heard the siren song in 1970 when he appeared on a BBC television programme about treasure-hunting. He took the reporter and camera crew to Epping Forest where he was searching at the time for a highwayman's hoard.

John didn't find that for the camera's eye but shrewdly made sure that they recorded some genuine finds without using miles of film stock. At 4 a.m. on the day of shooting, he slipped down to the forest and quickly detected some items, whose locations he marked with pegs. For the camera's benefit, he located the places again, then proceeded to dig up a variety of coins, musket balls and a few pieces of jewellery.

After his first television programme, John's life was blighted by a seemingly-endless stream of telephone calls, including an astonishing number in the wee small hours of the morning.

Thoughtless night-birds roused him from bed wanting him to describe in detail the principles for all metal detectors or let them have, then and there, a list of the best forty river sites. For peace of mind, John had his telephone disconnected; he has not summoned the courage to have another installed.

Not surprisingly, John's travels have implanted a map of Britain inside his head. He has visited so many places throughout the entire country that he could virtually trace a detailed outline of Britain's coastline without referring to an atlas. From burrowing through hundreds of books, he has also gathered much more than a routine knowledge of geology, gems, Roman, Anglo-Saxon, medieval and Victorian history, architecture, folklore and superstitions, numismatology, sociology, law, arts and crafts, weaponry, and military and naval history.

'In the beginning, I read about these things because I felt I had to,' he says. 'But the more I learned, the more fascinated I became about Britain and her heritage. And this, I believe, is the greatest indirect benefit you get from prospecting.

'It would be impossible to absorb so much information about Britain's history and not to care more about this country.

'Appreciating it more you can then understand its strengths and weaknesses. This is one of the things I would throw back at critics who are prepared to dismiss us out of hand.

'I would no more want to deface the countryside than I would want to harm my wife and children; it's all a matter of affection.'

The other professional is Tony Hammond, of Rustington in Sussex. (He is seen in action in Plate 10.) An alert and articulate man, Tony could strictly be called a 'drop-out', but only in a constructive way that is 180° from the usual meaning of that phrase. It would be more accurate to say that he 'dropped-in'. After years of envying people who enjoyed all aspects of their daily work, he chose, in 1972, to step off the treadmill of a successful business career to become a full-time treasure-hunter.

This took courage. Tony was past forty and had a wife and two daughters, then aged six and four, to support. In one simple action, he turned his back on a working life he had guided upwards through apprenticeship as a marine engineer, travelling and 'trouble-shooting' job for a London firm of heating engineers and directorship of a company making industrial heaters.

The catalyst for this dramatic decision was a simple BFO detector he bought for £20·00. A little work with this stirred some interest, which fermented when he saw a pulse-induction unit at an oceanography exhibition in Brighton. He bought one the same day and instantly realised its finding qualities. After several months of concentrated searching, he knew its money-making potential was strong enough for him to throw in his job. He began selling metal detectors under the trading name of Sussex Treasure Hunters, then became my agent for the Sussex region.

Tony sustains the point I made in Chapter 1 that anglers make very good treasure-hunters. He has lived all his life on the Sussex coast and was for many years a most successful angler. This gave him the intimate knowledge of local beaches that has been one of his greatest assets in his new trade.

John Webb is an all-rounder whose bailiwick covers the entire British Isles and Eire. By contrast, Tony has concentrated very largely on searching for rings along his beloved local beaches. Like John, however, he has free-lanced his services successfully to firms and individuals in need of expertise, mostly distraught housewives or couples who have lost an engagement or wedding ring. (True love, charting its usual erratic course, also boosts the sale of metal detectors, as it docs of flowers, chocolates, bandages and liniment. We have received a surprisingly large number of calls from engaged couples. They have had a tiff. Fiancée has wrenched off and flung engagement ring at her bickering betrothed. Anger abates and they make up. But they cannot locate the ring. So we have been asked to help. Often we have not only found the missing ring but also sold a detector.) However, coming back to Tony Hammond: his experience as a small-bore rifle shooter has attracted him to hunt for musket balls and ammunition, on which he has become an expert and historian.

Tony's decision to search mainly for rings was a shrewd one: he has something more than just an aptitude for it. In one four-hour session on a popular Sussex beach, he discovered twelve gold wedding and signet rings. The newspaper publicity he received after handing in this haul to the police prompted a married couple to write, asking if he would hunt for the wife's wedding ring, lost several months previously at Bognor Regis. It was a small miracle that he did so very quickly. During this search, he also dug another eight gold rings.

Tony Hammond advocates—as do many treasure-hunters—keeping the detector switched on while walking from one site to another; the unexpected often comes within range of the detector head. One day, *en route* to check a favourite glory-hole on a beach, he was startled to get a reading above an area he has searched several times previously with little success. He probed around and came up with a Victorian penny. During the next hour he found seventy coins, including one George III penny, and a mass of copper coins, ranging from Victorian to Edward VII. The signs were that tidal patterns at that spot had altered from their norm and washed the coins to that previously barren area. The pieces were badly corroded, so financial gain was negligible. The reward came from the lesson that no area of a beach which is subjected to tidal disturbances can be completely discounted for possible finds.

As a ring-finder, Tony has what the petrol commercials used to describe as 'miracle ingredient X'; he puts it down to an unfathomable sixth sense. Being aware of this, he is prepared, from time to time, to ignore a methodical and well-reasoned approach to a survey and head straight to a spot where his fancy has guided him. His exhaustive readings of tides and currents, his knowledge of people's accustomed movements on that beach have told him he should not really expect to come across anything worth while, or indeed anything at all, in that place. Yet very often the hunch, followed against the rational grain, has proved right.

Tony is at a loss to explain why he does this, just as some bowlers at cricket do not know why they can make the ball swing further or lift higher from the pitch than other bowlers. He quietly accepts that this talent resides within him, and that many other people also have it. So he counsels tyro treasure-hunters who feel compelled to stroll up to and work a specific location for no explainable reason to surrender to this instinct.

'It has paid off for me many times, so it should do so for other people,' he says.

'The electronic circuitry of detectors has made good treasure-hunting something of a science, but there are other powers, perhaps even mystifying powers, we can possess to find things. Let's use those powers.'

Tony's acceptance of occult forces in his chosen work has led him to bring into some of his prospecting the ancient practice of dowsing. To most people, this simply means water divining, which is, in fact, only one branch of dowsing. (The whole science of dowsing is too complex for me to go into fully. If you are interested in learning more, please consult your encyclopaedia, or seek out a dowser.)

One of the more spectacular aspects of dowsing is distant detection. This involves a pendulum. Before using it, an individual must find out whether he gives off positive impulses—which make the pendulum move clockwise—or negative impulses inducing anti-clockwise motion. Under traditional methods, the dowser holds a specimen of a sought-after substance (gold, bauxite, or whatever) or an object related to the thing he is seeking (if it is a woman's locket, he holds another of her possessions, a comb perhaps). The latest refinement of this technique dispenses with the sample; a clutched slip of paper bearing the sought article's name in block capitals will suffice.

The other hand dangles the pendulum over a map or plan of the area under study. Then the pendulum should gyrate towards the location of whatever you are looking for; the precision of its movements depending on the accuracy of the map or chart.

Tony says: 'It is obvious that a combination of dowsing and searching with a detector presents an exciting approach to treasure-hunting.

'I don't profess to be an expert, so I have sought the aid of a very experienced dowser with a number of notable successes to his credit.

'The most interesting part is that he is an ex-Sussex man now living in Australia. How's that for distant dowsing? From 12,000 miles away, he helps me to search for objects in Britain.'

His friend scored a notable *coup* in Australia with his dowsing. Using a snapshot of a woman, taken in the 1920s, he accurately pinpointed the location of a crashed aircraft in Australia in which the woman had died almost a half century ago. Several previous searches had failed to locate the wreckage.

This seemed uncanny, but then, so does Tony when he takes a detector on to a beach and locates specific articles lost months previously. The late Professor Tolkien beat us to the title, but we are happy to acknowledge that Tony Hammond is very much the contemporary 'Lord of the Rings'.

From two professionals, both absolutely content in their work, I'll move along to consider a young man who cares about prospecting the way some people care about their dogs or budgerigars: with an overwhelming passion.

His name is Tony White, and he comes from Surrey, whose hills, dales, streams, forests and ancient pathways he knows as intimately as any surveyor in the county. Tony happily makes the sacrifices of the true fanatic to take his detector out on searching expeditions.

For instance, his hours of sleep suffer. Almost every week-end of the year, and many weekdays as well, he is up at first light to speed off to a likely site. An outside fitter by trade, he drives a van to many job locations with his detector stretched along the rear window ledge. Tony is suffused by the lovely optimism that, wherever he goes, the find of a lifetime could be near by, so the detector must be at hand.

He often skips lunch: that hour is too valuable to waste when the coins and artifacts of twenty centuries of Surrey history are waiting to be discovered. Sometimes he can't suppress his impatience; he will search all morning, then catch up on his work in the evening; that seems the proper sequence to him.

Treasure-hunting brings out both the poet and dreamer in him. He will stand on a hillside and visualise a Roman legion tramping along a path snaking down into the valley, pennants stirring, armour clanking and dust spurting up from their sandals as it all happened on that very spot long ago. Surveying a landscape, he will blot out from his sight the existing features and try to imagine it as nothing more than a mass of gorse, scrub and tangled growth. This mental reconstruction helps to guide him to the hollows, the outcrops and other places where he thinks people used to gather in that area. This sixth sense works wonderfully well for Tony and he certainly needs it because he tends to ignore the slog of poring over books and other sources of reference. He has a special intuition that most other people don't have. So I don't recommend that newcomers copy his fanciful ways; most of them will have to bow to the sweat of routine research.

Tony's passion really sprang from the feeling for the history of the Surrey countryside that he felt as a youngster. When visiting his grandmother, who lived near Dorking, he would prowl fascinated along an old track reputedly used by opium smugglers. He would visit the site of old Roman villas, walking on tiptoe from fear that even his small footprints would damage something valuable.

Almost as soon as he bought his first BFO detector, the good finds started coming. He displayed a knack that suggested buried coins began to thrust their way up into his detector range when they sensed he was approaching. After one successful day (a fine haul of shears, a ploughshare, picnickers' knives and forks), he was climbing through a fence with the detector ticking, when he got a very small reading. He unearthed a perfect Roman coin, then four others. Back at five o'clock the next morning with small spade and sieve, he found eleven more coins, from the eras of Constantine I and II, among others.

Another day's treasure-hunting brought the haul of these thumbnail-sized coins to twenty-four, a marvellous début for someone who had scarcely begun to cope with the techniques of the business. Tony took them straight to Guildford Museum, whose curator was surprised and delighted; it was the first group of coins a member of the public had brought him.

Not long afterwards, Tony and a friend decided haphazardly to work a footpath running through a wood. They quickly found a cluster of fifty coins: one gold, plus groats, half-groats and pennies, two of which dated from the reign of Henry V. After the police returned it, Tony kept half of his collection because he couldn't bear to part with the coins.

'I don't believe in searching just for profit; the tingling feeling that good finds can bring you are enough for me. My wife and friends may, at times, call me a fanatic, but I prefer to think of myself as an interested party.'

Certainly his flow of finds has been enough to keep anyone interested: a William IV five shilling piece dated 1696, an 1821 sovereign in mint condition, a steady stream of copper Napoleons from a favourite site, arrowheads, brooches, snuff-boxes and artifacts, a small mountain of bottles and an assortment of rings, at least five of which still adorn his fingers.

While Tony's methods of choosing his working sites are unorthodox and perhaps unscientific, his techniques with the detector are certainly not.

> 'When I decide on a patch of likely ground, I work it at least four times, side to side, then up and down and coming across it from different angles.
>
> 'This is what you must do to find that hoard. The object may be so small it gives only a tiny beep on the detector that you could miss on one of your swings. On one strike it could have been just out of your range; come back along a different line and you have a good chance of detecting it.
>
> 'I've worked some sites three or four times. Then in the first minute of the next visit, I've found a coin just near the surface and wondered how I could possibly have missed it before.
>
> 'Some people who don't do so well say that I'm lucky, that I keep my interest up because I keep on finding things.
>
> 'But I come across my finds only because I'm patient; without patience you are lost.' (That says it all.)

Like a great many dedicated treasure-hunters, Tony prefers to work alone—without human company, that is. He often takes along his old English sheepdog, Humphrey, 'who'll second my opinion that it's a good hobby'.

Tony's wife, Caroline, stays at home pursuing her vital interest of looking after their three children. Caroline doesn't growl when he comes home late of a summer evening, perhaps tracking in sand and leaf mould from his shoes. When Tony brings home a sackful of disgustingly greasy and dirt-caked bottles, he'll find them next morning neatly arrayed and the cleaning process under way.

Tony appreciates this quiet sympathy for his fanaticism: he compares it to the latitude some other treasure-hunters are allowed.

Like a missionary inflamed with zeal, Tony has converted infidels from other recreational religions: one fishing friend has tossed away his rods, a former shooting nut has locked his shotgun into its rack. Both now stalk through the dawn mists with detectors. Their reformation delights Tony, but their wives have not been available for comment. The ladies shouldn't fret because their husbands and the reputation of treasure-hunting are in very good hands.

CHAPTER 12

ARCHAEOLOGISTS AND TREASURE-HUNTERS: THEIR STRAINED RELATIONS AND WAYS TO SOLVE THEM

WE all have nightmares. I suspect that Britain's small band of hard-working archaeologists suffer them more than most people. Every day of their working lives they have to confront a host of problems that they classify as menaces. These foes not only refuse to go away; they keep growing larger and larger to compound the archaeologists' sorrows.

So one can visualise the archaeologist tossing fretfully in his bed as harrowing dreams beset him. In his nightmare, he is a puny figure, menaced on all sides by an army of huge mechanical monsters. Lumbering towards him from one side is a giant bulldozer, its headlights blazing like the eyes of a fierce animal, its huge blade and tracks callously crushing everything in its path. Closing in from another direction is an immense tractor pulling a multitude of harrow prongs. Despite the archaeologist's cries, all these vicious metal tongs bite deep into the earth, throwing up splintered and destroyed shards of pottery, sacred bones and chunks of rubble that were once parts of Roman villas and medieval keeps. Also pressing down remorselessly on the tiny shrieking figure is a gargantuan cement mixer, clanking and groaning, spewing out an ocean of wet, grey slime that covers the land back to the horizon and threatens to swamp, to drown the petrified archaeologist.

These are terrifying visions. But the archaeologist has no respite from them; they are part of his waking hours as well as his tormented dreams. The cold, grey tide of hardening cement is thwarting his efforts to retrieve and analyse the artifacts that make up the jig-saw puzzle which is Britain's history. This important task has never been easy; there has never been enough money or personnel to do all that needs to be done. So our knowledge of the country's history is notoriously shallow, often mistaken and riddled with patches that are either blank or appallingly sketchy.

For instance, general beliefs about ancient Roman standards

of hygiene were almost completely wrong for two thousand years. Down through history came the glowing image of the Roman in crisp and spotless toga, his body freshly scrubbed and anointed with oils, his brushed-forward locks arrayed in shiny and greaseless curls across his forehead. Only recently we have learned that most Romans were like all peoples of their time: grubby in dress and personal habits. Even more recently the layers of rectitude that have encased the Victorians have been brutally wrenched aside to reveal the sweaty and licentious real people beneath them. Many of the things her subjects seized hold of when they laid their Bibles aside would have horrified their august monarch.

Archaeologists are the sleuths meticulously and tirelessly piecing together our heritage, filling in the misleading and bewildering gaps. They need and deserve all the help that the population can offer them. Their handicaps are many and their collective resources too few.

An antiquarian (he wished to remain anonymous) with an intimate knowledge of British archaeology, estimated that in early 1974 there were perhaps a total of five hundred professional archaeologists in the country. Assisting them fairly constantly or spasmodically were some 4,500–5,000 amateurs. The figures were by no means precise; another expert reckoned the professionals could call on the services of twice as many, i.e. some 10,000 willing helpers; a third said 20,000.

Whatever the accurate figure, archaeologists have been, year by year, losing ground both literally and figuratively. There are too many forces for change deployed against them: property developers, construction companies, the national and local governments, roadmakers, lobbyists, etc. The growing population has needed more and more land (a great deal of it of intrinsic historical interest) to house and feed it. Archaeologists have cited deeper ploughing methods (which cleave through vital layers of ground) and inexorable development works as their greatest worries.

To carry out what they choose to refer to automatically as 'progress', the officials and entrepreneurs have had tens of millions of pounds and hundreds of thousands of workers at their disposal. In this unequal struggle, the archaeologists have naturally stood no chance; they have been overwhelmed.

As mentioned before, archaeologists do not receive anywhere near enough money to tackle even a satisfying percentage of the work they have to do. The director of the Council for British

Archaeology, Mr Henry Cleere, said the Department of the Environment in 1973 allocated a total of £813,000 for all types of archaeological work. The lion's share of this money was spent by the Inspectorate of Ancient Monuments on excavation work. The Inspectorate's full-time officers and amateur helpers have had to concentrate on trying to search and analyse sites before the juggernaut of progress swallowed them up.

Mr Cleere said local government bodies throughout Britain supplemented the Department of Environment allocation with grants totalling approximately one million pounds. He said that the Council for British Archaeology received a grant of £17,500 from the Treasury. Another informant calculated archaeologists needed a minimum of £4 million to mark time; instead their collective anticipated revenues were less than £2 million.

Archaeologists in Britain have to contend with an equation that adds up to a massive headache for them: they are trying to retrieve and preserve the relics of a very long history on a small and over-populated land mass. An unofficial source considered that an average of 8,000 historical sites were lost in Britain each year, adding up to a shattering 80,000 in a decade. Even if this figure is halved, you can appreciate why archaeologists would be prey to those horrific dreams.

An official from the Inspectorate confided that 'an incredible amount of frustration' had built up. This had been tempered to some degree by the amount of successful and very important excavation work performed during the early 1970s. The department appreciated the immense outside help from RESCUE, which had done much excavation work, and from individual organisations. Lloyds Bank, for instance, had paid for excavations at two sites where it had planned to build new branches. Some developers had lent their machine plant for site clearance and made concessions for certain aspects of the excavation work.

However, these spurts of help have been random patches of light in a generally overcast picture for archaeologists. They have been appalled by the rate of loss and very testy towards those they branded as culprits. This is where we come in. High on their list of needless offenders are treasure-hunters. Archaeologists lash out vigorously at people who buy and use metal detectors for any purpose other than their own line of purely scientific inquiry.

Of course, many archaeologists use metal detectors on sites. They are quick to point out that such use is a minor part of their work. And besides, unlike their idea of the vast majority of treasure-hunters, they are informed people who know what, and what not, to do on sites. For years, archaeologists have generally been most reluctant to incorporate new scientific methods in their work.

Staunch and well-placed supporters have peppered the letters columns of national and local newspapers with irate letters. Many of these acidly dismissed all treasure-hunters as greed-crazed and totally ignorant vandals. Supporting these outbursts has been a flow of newspaper and magazine articles, some of them (not penned by professional archaeologists) most offensively phrased. In these the metal detector has been passed off as merely a 'toy' that yahoos use in the hope of quick riches; they are afflicted by 'complete ignorance' of archaeology; they are, to a man, nothing but 'fortune-seekers'.

A spokesman for archaeologists was once quoted along these lines in a newspaper story: 'They [treaure hunters] get a buzz, then dig a dirty big hole. At the end it looks as if a massive mole had gone berserk. It is estimated they have ruined a hundred archaeological sites.'

While one sympathises with the problems that archaeologists constantly face, some statements have been wildly overwrought. It is the inescapable truth that amateurs with metal detectors have wilfully trespassed on to restricted sites. It is also true that highly-placed bank officers have been convicted of embezzlement; that stockbrokers have been found guilty of unethical practices. Surely it is just as illogical to brand all treasure-hunters as vandals as it would be to make the sweeping statement that all bankers and stockbrokers are crooks.

If the spokesman's estimate of one hundred ruined sites was correct and each had been ravaged by a different person, that left well over 30,000 treasure-hunters who were not guilty of such atrocious behaviour. I must repeat my remark from the opening chapter: gross and deliberate intrusion on to archaeological sites is almost entirely the work of thieves using metal detectors for their aims.

Such people have no place in the mainstream of prospecting ranks; they are a small percentage of the numbers who have acquired detectors. All but a few persons who buy shotguns do so to indulge in some type of sport: clay-pigeon or skeet shooting, stalking various kinds of birds and game. The remaining handful obtain the weapons to hold up banks and armoured cars or dispose of their nagging spouses. The same applies to treasure-hunters: most have innocent intent. They deserve the benefit of the doubt; the prime assumption should be that they have purchased a metal detector to use *always* and *only* in legitimate ways.

Having said that, I must concede that critics of treasure-hunters are sometimes right. Some people do venture through ignorance on to archaeological sites and other restricted areas. Some dig holes and wander away without filling them in again. Some break the law by not declaring treasure-trove.

When society has become a perfect structure at all levels, it could theoretically produce perfect citizens. That day has not arrived and most likely never will. So we are forced to accept people's imperfections, while striving to keep their wrong, foolish and anti-social behaviour down to a minimum. Treasure-hunting is very much in its infancy. It has had no bulky compendium of Standing Orders, no traditions to draw on. As this book is written, no national supervisory body exists to lay down a mass of regulations or to co-ordinate the work of individual clubs. Such a national body is needed and, hopefully, will be set up before long.

In the meantime, we strive to guide newcomers along the right lines. Every person who buys a metal detector from us receives a copy of the Code of Conduct, printed in Chapter 2. This is far from perfect but does a very useful job. We make sure that each purchaser knows that he must take out a licence before using the detector.

If everyone obeys the code to the letter, the antipathy that sadly exists between archaeologists will diminish, but will not vanish. It may be too irreconcilable to be removed completely. So what can be done to reduce tensions, to bring both sides towards a state of accord where they know and understand each other's problems more clearly?

Firstly, it would pay to consider major reasons why a sense of conflict exists between some members of the two groups. It could boil down to the difference between their basic motives for activity. It has been described this way: 'Archaeologists excavate for retrieval and scientific analysis of items, many of which have high intrinsic but low monetary value. The feeling is that treasure-hunters search and dig for personal gain; this can be either to retain what they find or sell it for money. It is almost impossible to reconcile the two motives.'

This clash of ambitions has aborted some previous overtures towards peaceful collaboration. Mr Henry Cleere said that a group of treasure-hunters once wrote to the Sussex Archaeological Society, of which he is a member, asking if it could affiliate its members *en bloc* with the society.

'The treasure-hunters enclosed a copy of their Code of Conduct, which is a very good code as long as it is observed properly.

'The Archaeological Society wrote back to say it would need certain undertakings about the treasure-hunters accepting disciplines and acting under the instructions of qualified people.

'In reply, a letter came back stating a crucial phrase: that the treasure-hunters would be happy to assist on excavations, provided they could keep some of the articles they found.

'This goes against everything that archaeologists believe in; people cannot keep what they find; the items must be available for study and exhibition. Souvenir gathering must be kept in the clearest perspective.

'I have kept a Roman nail from a site in East Anglia, but since there must have been a million of them, this was no crippling loss.'

In this instance, quoted above, the treasure-hunters and archaeologists could not find enough common ground to amalgamate or pool their efforts.

Some critics of treasure-hunters oppose links of any kind. They consider these would confer on treasure-hunting a respectability that they do not feel would be warranted. Even relatively neutral observers contend that archaeologists should not yield, or soften, their critical stance. The reasoning here is that historical and scientific searching weighs far more heavily on the scale of values than the gratification of seeking articles for personal gain. Anyone who deliberates on the vast amount of historical data still imbedded beneath British soil, and the troubles that impinge on archaeologists from other directions, cannot dismiss, or belittle, the high degree of validity in this view-point.

However strong the archaeologists' case and the wretchedness of examples of site violation, every square inch of Britain is not part of the National Trust nor on the list of Scheduled Ancient Monuments, much as some zealots would like that to be so. As citizens, treasure-hunters have their civic rights. These cannot be wrenched away from them. The Department of Environment accepts this point. In 1973, the Council for British Archaeology asked the Department to draw up 'draconian' legislation. No details were released, but any such proposed law seemingly would have crushed the pastime of treasure-hunting. The Department refused the request.

Archaeologists are disturbed by what they consider to be the inadequacies of the existing regulations on treasure-trove, which stipulates that only objects of silver and gold belong to the Crown; when found they must be reported. They contend that Britain is unique in having no firm laws covering the finding of portable antiques. They term the treasure-trove regulations hopelessly archaic and riddled with dangerous loopholes. They have been pressing for years for stronger protection to cover all or most articles of historical significance.

In the meantime, they have attempted to deter intruders by omitting grid references from the list of Scheduled Ancient Monuments (which has swollen by well over 20,000 additions since 1971). This is a valuable reference source for treasure-hunters to study and note forbidden sites. Despite the flood of new entries, a great many archaeological sites in Britain are not sign-posted so the treasure-hunter, or any other visitor, unfamiliar with the district, cannot be aware that he has no legal access. This underlines the importance of trying, as far as it's possible, to determine who owns strange land before you venture on to it. As further protection on marked and unmarked sites, archaeologists are cited as having 'bugged' them by scattering around tin-tacks, ball-bearings, aluminium washers to confound metal detectors.

The enduring schism between the two groups was summed up by Professor Charles Thomas, M.A., F.S.A., in 1973, when he was President of the Council for British Archaeology: 'My council has failed completely to set any level on which this hobby and scientific archaeological research can meet.'

Other spokesmen have stated that their main complaint is not directed towards metal detectors; it is the subsequent digging that galls them. The trowel or spade is their major enemy because it slices through the layers they want left intact. An inspector from the Inspectorate of Ancient Monuments said he once plotted the pattern of an entire vanished building on an excavation site near York by carefully removing 3,000 iron nails.

'A person going indiscriminately over the site and picking up nails, would have lost the plan of that building completely,' he said.

Henry Cleere has maintained a more open mind than many. 'One would welcome a concerted approach by the various [treasure-hunting] bodies, some of which are attempting to achieve respectability, for a round-table discussion which would help us to resolve the conflict.

'We are always on the look-out for more information, from field-workers of all types. There are people who walk over ploughed fields, who study streams. They look at the things they find, the structures, the shape of the land and feed this information back to us.

'Ideally what we need is a mechanical or electronic device allowing us to look a little bit under the soil without actually disturbing the soil. This would be extremely valuable.

'My purely personal hope is that treasure-hunters will not just look upon their jolly little machines as an opportunity to get something for the mantelpiece, or to sell, which is even worse.

'I would like many of them at least to develop, to refine their techniques, to calibrate their machines in some way so that they will be able most times to know when they get a reading, that the hidden item is a coin, a piece of iron or whatever. They could record this information on to a plan or map. This information would be fed into a central archive, which will, I hope, be built up some time in the future.

'It would be available for subsequent investigation as and where it can be fitted into a local or national programme. We want people not to dig the holes but to use their skills or develop new skills and not to use their metal detectors in a blind way.'

For that round-table discussion to become feasible, i.e. acceptable to most archaeologists, they believe the compromises should stem from the treasure-hunters. As stated earlier, the archaeologists intend to stand firm in their attitudes.

Yet some adjustment by both parties will be needed. One experienced treasure-hunter commented: 'Archaeologists must realise that the man with the detector is here to stay. The treasure-hunter must appreciate that the archaeologist is a person to be respected and his cause assisted in every possible practical way. And there must be mutual agreement, some code whereby each knows and respects the other's boundaries.'

Some very succinct advice for treasure-hunters has come from the C.B.A.'s secretary, Miss B. de Cardi: 'Stick to river banks.'

This crisp statement implies that every possible location is *de facto* an archaeological site and the sole preserve of qualified members of that profession. It is a far narrower concept than others offered by staunch supporters of the archaeologists' view-point.

The antiquarian, who believes archaeologists should not budge one inch from their traditional stance, says they cannot object to any prospecting and digging on beaches, and on river foreshores and tidal estuaries. The irresistible movement of water down the centuries has long since disturbed, and often obliterated, pure archaeological evidence.

As I stated in the chapter on rivers, tidal rivers are the most prolific sources of finds. Also a vast number of beaches, intelligently worked, yield handsome returns. Treasure-hunters who who choose to search only these places can build fine collections and make money from selling their finds. But such rigorous self-denial would largely block them from obtaining worth-while quantities of immensely popular items, notably ancient bottles and pot lids and ammunition. They would also be denied a chance to look for hoards, which, under existing legislation, they have a right to seek. A onsiderable number of hoards would be buried in secluded places whose archaeological interest is negligible.

Everything, including causes as worthy as archaeology, must be kept in some kind of perspective.

As a veteran of his trade, John Webb says his numerous overtures to archaeologists for a round-table conference had come to nothing. He felt the increasing number of treasure-hunters' clubs around Britain played useful roles in making members aware of their responsibilities. But as separate units, they lacked the muscle and co-ordinated voice needed to try to smooth out the differences with archaelogists. For this job, a nation-wide club was needed but finance was a major obstacle to it being formed in the foreseeable future.

John Webb says: 'With so many archaeological sites going under the plough, excavator and drag-line every year, archaeologists need every volunteer they can recruit.

'Calls for volunteers to help on excavation sites have been almost a daily thing for a long while. They can't get all the help they want and there are more than 30,000 detector users, many of whom would help free of charge on a site for the sheer love of it under the instruction of qualified architects.

'There is no doubt that more sites are lost than need be because the full pool of volunteers isn't tapped.'

I wrote in the opening chapter that experience very often changes the attitudes and ambitions of treasure-hunters. This applies especially to a sense of history. A great many start out with a blurred and fragmentary idea of Britain's extraordinarily rich and dramatic past; a few dates, the names of one or two kings and queens can be virtually the sum of their knowledge.

But the solid research which is vital for successful all-round treasure-hunting brings history to life. It becomes much more than a string of arid and essentially isolated dates. The gaps are filled in. Perhaps for the first time, the researcher sees his country's history in terms of people, who are endlessly fascinating. He learns of the clothing and ornaments they wore for commerce, recreation and battle; the houses they lived in; the furnishings and bric-à-brac that filled those houses; the utensils and vessels they used for eating and drinking; where they caroused, gambled, frolicked, courted both love partners and business partners and how they betrayed both; how they survived or succumbed to pestilence and danger; how they slavishly followed or broke the laws of man and God.

As a result very many treasure-hunters are more alert to, and moved by, their heritage than their criticis may believe. They are not fools; they recognise historical vandalism as such; they are no more likely to indulge in it than the manager from the bank on the corner would yield to the temptation to run off with the money. It would not be beyond their talents and interests to work harmoniously with archaeologists as members of excavation teams.

Knowing this, we urge newcomers to treasure-hunting to acquaint themselves with the background, the aims and techniques of archaeology. Two recommended books for part of this study are Sir Mortimer Wheeler's *Piecing Together the Past* and Graham Webster's *Practical Archaeology*. You could also study in your reference library or buy from Her Majesty's Stationery Office the latest list of Scheduled Ancient Monuments. The C.B.A. also publishes a Calendar of Excavations in eight monthly parts, listing sites where volunteers are needed.

By all means get in touch with your local archaeological society. You can offer to work as a site volunteer, using your metal detector when and where it would be useful. Those treasure-hunters who became absorbed in the subject, and prefer to work full-time under the control and disciplines of an archaeological team, are free to do so. In this way, a positive response to an offer of help would prove a fine bonus for archaeology.

Those who wish their treasure-hunting to remain a solitary and independent pursuit should heed Mr Cleere's words that archaeologists are 'always on the look-out for more information'.

Make a habit of carrying a notebook. If you come across

anything with the look and feel of antiquity: scraps of pottery and metal, unusual nails, pieces of fashioned timber (either above or below the ground), jot down the exact location. Then notify your local archaeological society or write direct to the Council for British Archaeology, 8 St Andrew's Place, Regent's Park, London NW1 4LB. What you have found could be of archaeological interest. You may also feel inclined to offer some of the items you find to your local museum.

Above all, never trespass knowingly on to a scheduled excavation site. To do so is stupid, inconsiderate and, of paramount importance, it is against the law. British treasure-hunters are less hemmed in by restrictions and regulations than are those in virtually every other country. Some people antagonistic to the pastime have been canvassing steadily for the introduction of stiffer laws which would inevitably curtail your recreational activity. If those critics can amass enough information about irresponsible, destructive acts, they could achieve their ambition.

Remember that each time you go out with a metal detector, you are an ambassador for an activity that is rewarding in very many ways. Make sure that you face up to your responsibilities as both treasure-hunter and citizen. If you do, then the reputation and future of yourself and thousands of others who enjoy treasure-hunting will be secure.

I'm sure you have been told that before, but it's worth repeating.

It's in your hands to safeguard all those glittering prospects.

INDEX

Aberdeenshire, hoard found in, 49
Abersoch, beach at, 105
Aberystwyth, beach at, 102
ammunition, finding, 69–71
archaeology, 132–42
Aston Rowant, buried coins found at, 48
attics, investigating, 38
Auto-Pulse, 31

Balcombe, buried coins found at, 48
batteries, 32
beachcombing, 87–101; techniques of, 97–100; when, 91–3, 95; where, 93–5
beaches, information on, 102–12
Beachy Head, beach at, 102
beat-frequency machine, *see* BFO machine
Beaulieu Firth, 86
Beaumont, buried coins found at, 48
Beauworth, buried coins found at, 48
Bell, Alexander Graham, 23
Belstead, Iron Age treasure on building site at, 45
Berkshire, hoard found in, 51
Berwickshire, hoard found in, 50
Besch, Fred, 47
BFO machine, 22, 24–7; batteries of, 32; working of, 25
Bideford Bay, beaches in, 111
Blackfriars Bridge, 84
Blackhills, buried coins found at, 49
Blackpool, beach at, 103
Bognor Regis, beach at, 103
Bolton Percy, buried coins found at, 45
Borscar, buried coins found at, 49
Borth, buried coins found at, 49
bottle-collecting, 113–18
bottle dumps, history of, 115–16
bottle hunting, tools for, 116–17
Bournemouth, beach at, 103–4
Boyton, buried coins found at, 49
Bracklesham Bay, beach at, 104
Bredgar, buried coins found at, 49
Brighton, beach at, 104
buckets, glass-bottomed, use of, 14
Burntisland, beach at, 109

Caister-on-Sea, beach at, 104
Camber, beach at, 104–5
Canterbury, buried coins found at, 49
Cardi, Miss B. de, 139
Cardigan Bay, beaches in, 105
Cardiganshire, hoard found in, 49
Carsphairn, buried coins found at, 49
Chanctonbury, buried coins found at, 49
Cheshire, hoards found in, 44, 46, 50
Chesil Beach, 105–6
Chester, buried coins found at, 50
cleaning: bottles, 117–18; finds, 18–19
Cleere, Henry, 134, 136–7, 138–9
Cleuchhead, buried coins found at, 50
Closeburn, buried coins found at, 50

clothing, 16–17
Colchester, buried coins found at, 46, 50
Coldingham, buried coins found at, 50
commons, searching, 63, 65
conduct, code of, 20–1
Congressbury, buried coins found at, 50
containers, 16
Corriemonie, buried coins found at, 50
Council for British Archaeology, 133–4, 137, 142
Coyle, Roger, 47
Criccieth, beach at, 105
Cromer, beach at, 106
Crossland, John, 45
Cumberland, hoard found in, 48
currents, sea, 89–90

demolition sites, investigating, 39–41
Denbighshire, hoard found in, 51
detector: buying a, 10–11, 22–34; skill in using a, 63–5; the earliest, 23
digging: care in, 65–6; use of tools, 11
Disley, money found at, 46
Douglas, beach at, 108
Dove, Heinrich William, 23
dowsing, 127–8
drift, 26, 28
Dumfriesshire, hoards found in, 49, 50
Dymchurch, beach at, 106

East Runton Gap, 106
Eccleshill, coins found at, 45
Eclipse poison bottle, 114
equipment, 10–18
Essex, hoards found in, 46, 50

farms, searching, 61–2
Fifeshire, hoard found in, 51
Flamborough Head, beach at, 106
floorboards, 38
Folkestone, beach at, 106–7
footpaths, investigating, 66–8
fords, searching, 83
Forrest, Christopher, 44
Forth Bridge, 85–6
Foster, Eric, 30, 31
Fulham, 85

gardens, searching, 39
ginger beer bottles, 115
Graham, Peter, 45
Great Yarmouth beach, 93, 104
Greenwich, 84–5
Guisborough, buried coins found at, 51

Halsall, buried coins found at, 51
Hammond, Tony, 31, 125–8
Hampshire, hoards found in, 44–5, 48
Hayling Island, beach at, 107
Helmsdale, beach at, 107–8
Hertford, buried coins found at, 46
Hertfordshire, hoard found in, 46
Hindostan, 110
hoards: descriptions and locations of, 44–53; finding long-concealed, 54–61
houses, searching, 35–9

IB machine, 22, 24, 27–9; batteries for, 32; widescan model, 29; working of, 28–9
induction balance machine, *see* IB machine
Inverness-shire, hoard found in, 50
Ipswich: buried coins found at, 51; buried torc found at, 45
Isle of Man, beaches on, 108–9

John, treasure of King, 52–3
Jones, Revd Stanley, 44

Kent, hoards found in, 49
Kinghorn: beach at, 109; buried coins found at, 51

Kirkcudbrightshire, hoard found in, 49

Lambeth Bridge, 84
Lancashire, hoards found in, 46, 51
licences, 19
Lincoln, buried coins found in, 47
Lincolnshire: hoard found in, 47; King John's treasure in, 52–3
lines and pins, use of, 13, 78
Lizard Point, beach at, 109
Llanarmon, buried coins found at, 51
locks, coins hidden in, 36
London Bridge, 85

magnets, use of, 18, 78
Margate, beach at, 110
Moelfre, beach at, 110–11
Money Head, beach at, 111
Morfa Bychan, beach at, 105

Nelson, swords belonging to Lord, 47–8
Newbury, buried coins found at, 51
Newstead Abbey, buried coins found at, 46
Northampton, buried coins found at, 51
Northamptonshire, hoard found in, 51
Nottinghamshire, hoards found in, 46, 47

Ordnance Survey maps, 17, 66
oscillators, 25–6

permission, seeking, 69, 117
'phantom presence', 33
PI machine, 24, 29–32; Auto-Pulse, 31; batteries for, 32; C400, 31; working of, 30
Piecing Together the Past (Wheeler), 141
Pipe Locator's Licence, 19
ponds, searching, 68–9
Portledge, beach at, 111
Practical Archaeology (Webster), 141
Prestatyn, beach at, 111–12
prospecting, definition of, 7
pubs, searching rivers beside, 83
pulse-induction machine, *see* PI machine

Radley, Joseph, 54–6
rakes, use of, 13–14
Ramsbury, buried coins found at, 46
Rhyl, beach at, 111–12
rivers, searching in, 72–86
Roxburghshire, hoard found in, 50
Royal Charter, 110

screwdrivers, use of, 14–15
Selsey Bill, beach at, 112
Sheerness, buried coins found at, 46
Sheppey, Isle of, buried coins found on, 46
sieves, use of, 12–13, 80
Somerset, hoard found in, 50
Southwark Bridge, 85
stools, use of, 18
Suffolk, hoards found in, 45, 51
Sussex, hoards found in, 48, 49

Tadley, coins found at, 44–5
Thames, River: searching the, 72, 84–5; treasure found in, 47
Thomas, Prof. Charles, 138
tides, 74, 89
T/R machine, 24, 32
transmitter/receiver machine, *see* T/R machine
treasure-hunters: criticism of, 135–6; portraits of three, 119–31
treasure-trove, 20, 21, 138
Tricker, Malcolm, 45

Universal Magazine of Knowledge and Pleasure for 1784, The, 54

Vauxhall Bridge, 84

Wapping Old Stairs, 84
Warsop, buried coins found at, 47
Watchet Harbour, 112
water divining, *see* dowsing
Webb, John, 108, 120–5, 140
Westward Ho, beach at, 111
Wey, River, sword found in, 47
Whalley, Bronze Age treasure found at, 46
White, Tony, 128–31
Wick, beaches at, 108
Wiltshire, hoards found in, 46, 49
woods, searching, 59–61
Wrecks, Receiver of, 100, 110
Wybunbury, hoards found in church at, 44

York, buried coins found in, 51
Yorkshire, hoards found in, 45, 51